Created in China

Created in China

How China is becoming a global innovator

Georges Haour and Max von Zedtwitz

Bloomsbury Information
An imprint of Bloomsbury Publishing Plc

B L O O M S B U R Y
LONDON · OXFORD · NEW YORK · NEW DELHI · SYDNEY

Bloomsbury Information

An imprint of Bloomsbury Publishing Plc

50 Bedford Square	1385 Broadway
London	New York
WC1B 3DP	NY 10018
UK	USA

www.bloomsbury.com

BLOOMSBURY and the Diana logo are trademarks of Bloomsbury Publishing Plc

First published 2016

British Library Cataloguing-in-Publication Data
A catalogue record for this book is available from the British Library.

ISBN: HB: 978-1-4729-2513-8
ePDF: 978-1-4729-2515-2
ePub: 978-1-4729-2514-5

Library of Congress Cataloging-in-Publication Data
A catalog record for this book is available from the Library of Congress.

Typeset by RefineCatch Limited, Bungay, Suffolk
Printed and bound India

To Oscar, Lucas, Léonard, Charlotte & Céleste

To Emily, Cosima & Carl

Contents

Foreword

Never before in history has such a large country grown economically as fast as China has in the last 30 years. In this process, hundreds of millions of people have been pulled out of poverty. Such a size and rate of change come with particular characteristics. Here are some of them:

- More than 650 million Chinese people are connected to the Internet. Most of these are in cities and are remarkably engaged with hand-held devices and internet services.
- In recent years, wages have increased 8% per year on average. By and large, China is not a low-cost country any more.
- Among the world's 50 largest firms (by revenue), in 2016, seven are expected to be Chinese, all of them essentially in the oil and banking sectors.
- In order to develop domestic manufacturers of cars and planes, China founded a unique breed of firms, the 'corporate start ups', such as Qoros, for automobiles, and Comac for commercial airplanes.
- Such relentless economic development comes with substantial cost, including the environment. Ecological disasters are looming.
- In building its transport infrastructure, China is the world of superlatives, building the world's largest bridge, developing a network of 20,000 km of high-speed train tracks, as well as the world's highest airport, in Daocheng, Sichuan, at 4,410 meters. In the last ten years, China has invested 8.5% of its GDP in infrastructure each year, as compared with 2.6% for the European Union and 3.9% for India.
- In November 2013, the plenum of the communist party announced a continuation of the reform agenda, particularly in the financial and fiscal spheres. Related to this was the bold announcement of a new, large free trade zone in Shanghai.

No one has a comprehensive picture of this fast-changing country. This book intends to inform the debate as to whether or not China will be successful in transforming the 'manufacturing plant of the world' into a major source of innovation.

We document the patterns of this journey, with a particular focus on firms, as they constitute the essential actors of wealth creation. In the following chapters, we present our case for the positive term of the alternative, which, we believe, is not only good for China, but also good for the world.

Geneva and Shanghai, 2015

Chapter 1

Overview

The focus of this book is on companies, the key agents for generating wealth, moving China towards being a global innovator. As a new world actor of wealth creation, China must behave responsibly. It is hoped that the ethnocentric West will not repeat the grotesque brutality and incompetence of the opium wars in the nineteenth century.

The saga of modern China is quite out of the ordinary because of the scale and speed of change. Upon its creation in 1949, the People's Republic of China (PRC) built on a legacy of a remarkable culture, crafted over the course of fifty centuries of rich history. Then there was the disastrous rupture with ten years of violence and famine during the Cultural Revolution. However, in the last thirty years or so, the country has taken several hundreds of millions of people out of poverty. This remarkable achievement comes with three substantial liabilities: huge disparities of income, great damage to the environment and potential socio-political instability.

In many ways, China is following the path of Japan and South Korea although it initially relied on massive direct foreign investments, as well as cheap labour. A key step was China joining the World Trade Organisation in 2001. In the recent past, the focus has turned to stronger internal demand, home-grown developments ('indigenous innovation') and the internationalizing of Chinese firms.

Reflecting the magnitude of the country's economy, there is a sort of a Chinese theorem: on the one hand, commodities bought by China go up in price, because the demand is so large. On the other hand, the products sold by China force prices down. China is currently the world's manufacturing powerhouse, accounting for more than two-thirds of the output in DVD players, children's toys, and microwave ovens, as well as half of all personal computers/digital cameras.

China, however, is not just a (relatively) low-cost manufacturing country. Much of its manufacturing is moving to lower-cost countries, such as Vietnam. The Chinese economy thus captures only a small fraction of the value added. As a result, it is transitioning towards more value creation. Indeed, China's pragmatic and entrepreneurial spirit, massive investment in R&D, compounded by its Confucian tradition and the extensive use of the Internet by its urban population, mean that the country is about to become a major country for innovation. Part of this transition is the shift in the flow of investments. For the first time in 2014, outbound international investments from China were at the same level as foreign inbound investments into China.

The present book looks primarily at companies, but also, of course, at the environment in which they operate. In this momentous transition from 'made in China' to 'created in China', the following chapters describe the patterns of this remarkable shift.

Chapter 2 evokes China's brilliant history of its creative past. In addition to remarkable achievements in the arts, handicrafts and architecture, China is sometimes credited to 'have invented everything,' from printing to gun powder. It did not, however, turn these inventions into sources of wealth or power. The chapter discusses the possible reasons for this absence of 'commercialising technology.' This is in sharp contrast to the strong intent of contemporary China to achieve innovation-led growth, as portrayed by the indicators discussed in the following chapter.

Chapter 3 looks at the metrics, traditionally used as indicators to evaluate how innovation-intensive a country is. The uncertain nature of innovation makes it impossible to have reliable metrics measuring the impact and commercial success of innovation before it happens. Successful innovation is essentially a matter of quality of output, rather than quantity of inputs.

Input measures are used, such as the amount of R&D investments, in the case of technology-intensive innovations. Innovation, however, encompasses much more than those in the technical field.

Going behind the statistics, arguments in the chapter debunk some of the indicators, which are too often taken at face value. As we look at the highly complex process of turning novel ideas into commercial success, the discussion brings corrective elements (e.g. on patent output) taking into account the specifics of the Chinese situation.

Chapter 4 focuses on the key actors of the innovation scene in China: the companies, essential agents for wealth-creation, on the one hand, and the powerful, vertical government apparatus, on the other hand.

A remarkable feature is that the government is strongly and relentlessly committed to innovation. The latter fairly unilaterally defines the rules and the 'framework conditions,' in which the firms operate.

Chapter 5 looks at the country's framework conditions, which provide a context for innovation to flourish – or not. These include the tax code, the legal system, particularly the legislation on intellectual property rights, as well as the infrastructure and the financing of innovation in China. This environment is diverse, as it varies from province to province, continually experimenting and rapidly evolving.

Chapter 6 concentrates on the most crucial element for the success of the innovation process: the human factor. Not specific to China, the single most important ingredient for effective innovation is constituted by the talent and motivation of the staff involved. The Chinese scene presents several specific characteristics in this area, including the role of its large Chinese diaspora.

Chapter 7 documents the contribution of international corporations to China. Attracted by the 'world's largest and fastest growing market,' multinational companies from Europe, USA and Japan, but also Korea, have invested massively in the country. By doing so, they have contributed much more than investments. They have brought important know how in business practice and management, and have opened the country up to the non-Chinese world.

Chapter 8 describes the patterns, anticipated to be followed by China's government, in order to foster the development of a vibrant innovation scene. This concerns the ways government shapes the framework conditions for innovation in the country, but also the innovative ways occasionally manifested in public policies and practices.

Chapter 9 concentrates on the patterns of innovation expected to be followed by Chinese firms, as they increasingly become innovative enterprises. As they follow these patterns, in a fiercely competitive environment, companies will sharpen their skills for innovation-led growth.

The West has much to learn from the way China innovates. The country is already a place of innovation for business models, notably in areas involving the Internet and telecommunications. Chinese entrepreneurs are particularly well equipped to seize the benefits of the digital revolution.

Barring major mishaps in the country, this wave of innovation and entrepreneurial spirit will spread to many areas of social and economic activity. This is good for China and, if the world elects to handle this challenge in a positive way, this is good for the world.

Chapter 2

China's glorious history of innovation

It was Joseph Needham who, in 1969, articulated the question China had struggled with for over a century: Why did modern science and civilisation develop in Europe but not in China, despite all of China's past accomplishments? Until the fourteenth and fifteenth centuries, China and the West – synonymous with Europe and the other major civilisations of the time, the Islam and India – were on par. Europe had actually lagged behind, coming out of the so-called 'dark ages,' which lasted for centuries, following the collapse of the Western Roman Empire, while Islam maintained and developed further many of the science and philosophy of the antique world. China, on the other hand, had invented and used technologies, such as steel making, had developed a high division of labor and an economy that was more market-oriented than its European contemporary, and most of the land was privately owned. For all it mattered, the Industrial Revolution should have happened in China in the fourteenth century, not in Europe in the eighteenth and nineteenth centuries!

Talking about something that did NOT happen is highly speculative, especially when the matter is as complex as this one. However, scholars of various disciplines, including history, philosophy, linguistics, and Joseph Needham himself, have proposed theories why China failed where Europe succeeded.

The four great inventions

Concluding that the Chinese were 'just not inventive enough' is simultaneously tautological and wrong. Francis Bacon, one of Europe's most important scientists of the last 500 years, considered gunpowder, the magnetic compass, paper and printing as the four most important inventions that helped Europe's transformation from the dark Middle

Ages to the modern world. Ironically, all four of these inventions were made by the Chinese much earlier than they were invented by – or introduced to – the Europeans.

Paper is said to have been invented by court eunuch Cai Lun in about 100 CE, but it was already in use before, probably as early as 200 BC. Paper was important not only as a much more convenient medium for the written word; it was also used as the world's first paper money (i.e. banknotes). While other peoples still used papyrus, animal hide or stone tablets, the Chinese had access to a very light and durable form of communication and record keeping. The use of paper is intrinsically linked to book-printing technology, famously attributed to Johannes Gutenberg in 1439, but it was, in fact, used by the Chinese at least 400 years earlier; more than 700 years, if allowing for woodblock printing. Contrary to popular myth, gunpowder was used in China not only for fireworks but also for straightforward military applications. As early as the tenth century, gunpowder was used in incendiary bombs launched from catapults, and, even if their explosive power wasn't as deadly as it is today, the blast certainly scared the horses of an attacking cavalry and rendered their threat less potent. The north–south orientation in loadstones was already used for geomancy in China during the Han dynasty, but its application in magnetic compasses for navigation in the eleventh century was still 200 years earlier than in Europe.

In fact, one of the great accomplishments of Needham was the compilation of every mechanical invention and theoretical idea that had occurred in China up until modern times. This list included technologies such as cast iron, the stirrup, clockwork escapements, the wheelbarrow, matches, silk, and the crossbow. All of these are major inventions in their own right, with far-reaching consequences for society.

As an example, the stirrup was originally a loop suspended from a horse's saddle allowing the rider to mount the horse. Sometime in the third century the Chinese had advanced iron-making to the extent that they were able to make stirrups strong and durable enough to withstand repeated stress from riders standing in them. This led to a quantum leap in the use of cavalry for combat purposes as without this type of stirrup it is nearly impossible to fight on horseback as at least one hand is needed to stabilise the riders' position in the saddle.

This technology was adopted by the Avars who famously invaded Europe in the sixth and seventh centuries ('Avar' is their Turkish name: their original name was 'Juan-Juan' from Western China). Their ability to fight on a horse, shooting arrows and engaging in direct combat, gave them a tactical advantage over the defending Roman foot soldiers. In his Strategikon published in 580, Maurice Tiberius recommended the immediate adoption of stirrups made from cast rather than wrought iron – previously, horses were used mostly for transportation, not for warfare. However, only the Chinese possessed the technology at the time to make iron of the necessary quality; the Europeans still lacked the metallurgic expertise and furnace technology. Only after its introduction in the ninth century were European warriors able to fight effectively from horseback using an arsenal of weapons: the class of the knights emerged, and with it the advent of the feudal civilisation which characterised Europe's reemergence from the 'dark ages.'

The failure to capitalise on inventions – the isolation hypothesis

So, why did Europe, and why didn't China? While this may seem an academic question, 500 years after the fact, its answer begets a much more pressing question: Are the conditions responsible for China's shortfall in the fifteenth century still present in today's twenty-first century? Is China, despite all economic and political resurgence, in danger of missing out on the next transformative revolution? Has China changed at all?

Let us consider some of the reasons brought forward for China's failure to advance science and technology the way Europe did. One reason claims that China was, as a single empire, politically stagnant. One possible reason was that some of the inventions were made by artisans, who were not highly regarded by the bureaucracy of the Empire. Other inventions were made by scientists, who were perceived to be a challenge to the status quo, and were therefore rejected. As we know, innovation needs the leadership of a champion to succeed.

Another possible reason was geographic isolation. China was indeed difficult to attack militarily. China has a long coastline to the Pacific

Ocean. The examples of Great Britain, Japan and even the Allied landing in the Normandy during World War Two are testimony for how difficult it is to invade a country by sea. To the south, China was protected by heavy jungle, making it cumbersome for armies to pass through. The Himalayas block any attempt from the west, as do the huge deserts in the northwest and the chilling tundra in the north. There are few geographic barriers within China, making it easier for an incumbent power to remain in control. Once established, a government has little interest in seeing local innovations undermine its source of power; indeed the Chinese Government kept its scholars and scientists in constant check, just in case they posed a threat to its supremacy. However, the same cannot be said of China's neighbours, where new elites formed powerful armies of warriors. China was successfully invaded several times, most recently by the Mongols and then the Manchurians, and over its 3,000 year history, had plenty of civil war and various internal factions fighting for power. If war is indeed 'the mother of invention,' China had plenty of opportunities to be innovative.

Joseph Needham's 'Science and Civilization in China'

Science and Civilization in China (commonly abbreviated as SCC) is a series of books written and edited by British sinologist Joseph Needham on the history of science and technology in China. The underlying data and material is so monumental that even though Needham started the effort as far back as 1954, the series is not yet concluded and more books are planned. The series count currently stands at 27 books.

Science in China is actually Needham's second career. He was a biochemist originally, teaching at Cambridge University, and elected a Fellow of the Royal Society in 1931, whose 1939 compendium on morphogenesis was considered to become his *magnum opus* of scientific life achievement. But in 1937, when he was 37 years old, he came into contact with Chinese scholars who ignited his interest in China generally, and specifically its history of science and technology. He was stationed in Chongqing from 1942–1946 as the director of the Sino-British Cooperation Office. He used his travels to collect historic and scientific books wherever he could. He also met several Chinese scientists who were enamored with his devotion to Chinese history and science, and advised him on where to look for more. In 1946 he returned to Europe, initially to Paris where he became the first director of the natural science section of UNESCO, and then, as of 1948, back at Cambridge.

While still teaching biochemistry he devoted his attention to the SSC project. By 1948 the British Air Force had shipped his thousands of books on China from Chongqing to England. Taking stock of the material, he envisioned six books completed in a few years' time. However, in addition to his Chinese colleagues in England, he had even more supporters in China. They came from almost every scientific discipline, shipping yet several more thousands of books to him. For instance, meteorologist Zhu Kezhen sent crates of books to him in Cambridge, including the 2,000 volumes of the *Gujin Tushu Jicheng* encyclopaedia, a comprehensive record of China's past. Others used diplomatic channels when public mail was unreliable. They smuggled them past enemy armies and the police during times of war and civil unrest. All of these books now form the nucleus of the Needham Research Institute, which continues the work of Needham long after his death in 1995.

The 'isolation hypothesis' is also not a particularly strong one. It might have worked for Japan, which secluded itself for a century and a half, but China was always too accessible for traders and travelers. The Silk Road, which was a network of trading routes between China and the Middle East, provided ample opportunity for ideas and technology to be passed on to China. It certainly was a sufficiently reliable way for many Chinese products and technologies to trickle to the Levant and eventually into Europe. Marco Polo was not the only European to come to China; Jesuit missionaries, Japanese pirates and European navies also arrived. During the Ming dynasty (1368–1644), China successfully fought off Portuguese and Dutch fleets attempting to gain a foothold in the Far East.

In the early fifteenth century, Zheng He's 'expeditionary fleet' circumnavigated Cape Horn in Africa and possibly sailed as far as East Africa. This impressive undertaking was both military and representative in nature, expanding China's political and economic zone far into the

Indian Ocean. Despite some claims Zheng He did not reach the Western shores of North America, few doubt that he could have, had he wanted to. Under Kublai Khan, Beijing set up an Islamic Astronomical Bureau in Beijing which operated there for centuries. It used the mathematical and astronomical knowledge of the time and could have been a gateway to Persian, Arab and Greek science and technology. However, there is little evidence that these ideas were adopted by the Chinese. It seems that whatever foreign science or culture was presented to the Chinese emperors, it was not considered worthy of pursuit by the Chinese intellectuals.

By the eighteenth century, the British had become the strongest trading force in Asia. When the East Indian Company petitioned the Chinese Qing emperor Qianlong to open up to more trade, it resulted in the rejection that is still used today to illustrate China's complacent and self-centred attitude at that time. China was self-sufficient, had everything it needed, and was on top of the world. Being the Middle Kingdom, everybody else was a barbarian. Why challenge the status quo with foreign technology and ideas?

During the last 1,000 years, China has seen its fair share of foreign invasions and civil wars. Its relatively high standard of living and sophisticated culture attracted both domestic and foreign powers trying to gain a part of it. This explains (a) why China insisted so much and for so long on maintaining social harmony (and still does today under the current Communist Government); (b) why it developed a certain xenophobism over the years, especially with the rise of neo-Confucianism in the fifteenth century; and (c) why it established a form of administration and bureaucracy that was so efficient that every new emperor made use of it to control the huge population. The underlying Mandarin system focused on developing skills other than mathematics, engineering and technology, thus siphoning people away who might otherwise have become scientists and entrepreneurs, challenging the boundaries of technology.

In fact, China was so successful, that one could say that it really didn't need much innovation. Qiaolong's response to King George III may have sounded arrogant, but from the point of view of China, it is perfectly understandable. Historian Mark Elvin went as far as to suggest that China did not develop its own industrial revolution precisely because of its wealth, stability, and high level of scientific achievement. In what is known as the high-level equilibrium trap, Elvin claimed in 1972 that pre-industrial China had reached an equilibrium point where supply and demand were well balanced, production methods and trade networks were efficient and there was an ample supply of cheap labour, all of which resulted in poor incentives to invest time and capital in developing more efficient production technologies. During the Tang Dynasty (618–906 AD) China had been particularly innovative; many new technologies were introduced that allowed the economy and the population to be more productive with the result that they spent more time and effort on the arts and sciences. It is estimated that in 1750, Asia (with China its main economy) contained about two-thirds of the world's population and produced about 80% of global economic output, whereas Europe and America together accounted for only about one-fifth of the population and 20% of global economic output. The West was very much a marginal phenomenon at the time, especially from the point of view of China.

However, Chinese science developed without solid scientific theory, as noted in Toby Huff's excellent comparison of science in China, the Arab World, and Europe. Its lack of consistent systematic treatment of ideas and theories without unifying principles and methods was lamented even by Chinese scholars at the time. Scientific discoveries and breakthroughs continued to be made, but they were isolated and were difficult to build upon. Collected works such as Wang Zhen's Nongshu in 1313 were a highlight but also an exception of science during dynasties following the Tang (after 906 AD).

Wang Zhen's 'Nong Shu'

In the early 1300s, Wang Zhen was an official in the Anhui and Jiangxi provinces during the Yuan Dynasty. The creator of a wooden movable type printing system, in 1313 he published a treatise called 'Nong Shu,' a vast recording and collection of technologies used in China at the time. To some extent, Wang Zhen was both a Johannes Gutenberg and a Leonardo Da Vinci.

Although labeled a book on agriculture (the 'Nong' in Nong Shu means farming, agriculture), Wang Zhen's work really outlined the use of the various Chinese sciences, technology, and engineering practices, going far beyond farming applications. It is a treatise on the state of Chinese science and technology of the day, covering hundreds of Chinese inventions in many different categories (e.g. hydraulics, water powered bellows, the bamboo water wheel, and the hydraulic drop hammer). He illustrated them carefully so that it was easier for his readers to replicate them. One of the reasons why he wrote the book was to help the destitute and impoverished farmers who had suffered from the oppression of the new Mongol rulers. With more than 110,000 characters the book was exceptionally long – it was not meant to be read by the farmers themselves (who were mostly illiterate) but by local county officials who were in charge of improving both agricultural productivity and economic livelihoods.

Wang Zhen himself was an innovator, inventing a technique that made it easier to handle movable type pieces in wooden type printing. He described his invention in an appendix chapter in Nong Shu. Although 'Nong Shu' was not printed using Wang Zhen's technique, his innovation was useful in spawning more movable wood type printing technologies that were used in China until the introduction of the European printing press in the eighteenth century.

Taoism and Confucianism

This dismal state of fundamental scientific thought coincided with a shift of espoused philosophy from Taoism to Confucianism around the fourteenth century. While the Taoist paradigm had promoted, in its own way, some form of mathematical and scientific exploration of nature, the Confucian paradigm changed the focus to social philosophy and morality. At the same time, Europe-based Western science and philosophy, which to a great extent had been harboured by the Church, transitioned from the scholastic period into the Renaissance, essentially moving in the opposite direction. The Renaissance was Europe's scientific revolution, preparing the groundwork for the industrial revolution 200 years later.

Unfortunately, neither Taoism nor Confucianism appear to be philosophies conducive to innovation and technological progress. One of the most influential proponents of Taoism was Zhuang Zhou, a near contemporary of Confucius living in the fourth century BC. A sceptic, his aversion to abstraction and analysis is well documented: 'Those who divide cannot see,' he writes in a book bearing his name, diametrically opposed to what was later to become the underpinning theory for Western science: reductionism. Taoism also encompasses the School of Yin and Yang, which is emblematic in the Western world for its holistic and harmonious integration of potentially conflicting opposites. It is difficult to envision Western science, with its absolute statements about right and wrong, to function within such a philosophical system.

Confucianism became China's first state ideology right after the downfall of China's first emperor, Qin Shi, succeeding a brief but important period of legalism or rule by law, itself the ideological predecessor of the Mandarin system. Confucianism focuses on ethics, morality and relationships among people. One of its central tenets is the relationship between the superior (ultimately, the emperor) and his subordinates – an ideal philosophy for running a country with a huge and potentially unruly population. In its strictly hierarchical system, everything is fixed. Students have to respect their teachers, wives their husbands, younger siblings their older siblings. There is no room for external change or foreign ideas. In fact, one of the reasons why China still has such an ambivalent relationship with Westerners today is precisely because Confucianism is silent about the relationship between the familiar and the foreign. Respect for tradition discourages new developments in favour of maintaining harmony and stability. Thus, already more than two thousand years ago, anything that challenged the supreme, the central authority, had to be rejected, and that included innovations that moved power into the hands of the emperor's subjects.

Life as a philosopher, scientist or scholar used to be dangerous in China. In an attempt to unify thought and political opinion, in 213 BC Emperor Qin Shi Huang ordered the burning of many philosophical, scientific and political books. He is also said to have buried alive hundreds of scholars and scientists in the clean-up following the burning. Whether or not they were actually buried alive, many of them were killed; this set an example for future generations of scientists to be careful about expressing their thoughts and opinions. These events had a particularly devastating effect on one of the most promising schools of philosophy at the time, the Later Mohists. They pursued a form of scientific thought and argumentation which most closely resembled those ideas proposed by the ancient Greeks, the progenitors of Western science.

Oppression of thought and discourse is, of course, not particular to China. Neither is the large-scale extermination of people with different belief systems or religion. Unfortunately, all that made China so strong and powerful also made it very efficient in eradicating non-conformists and disbelievers.

Role of the script

One of these common systems that everybody in the expanding boundaries of China had to adopt was the Chinese script. Heralded by Western philosophers such as Johann Wolfgang von Goethe as a universal language-independent medium of communication, it has also been singled out as a barrier to scientific progress and innovation by those endorsing the Alphabet effect. According to this theory, a greater level of abstraction is required due to the greater economy of symbols in alphabetic systems; this abstraction and use of the alphabet coincides with abstract science, deductive logic, codified law, individualism and innovation in general. In Chinese, all thoughts, old and new, must be written in one of the thousands of accepted Chinese Hanzi characters. New terms cannot be created easily; these are usually constructed by combining two accepted characters, or by reassigning a new (or additional) meaning to an existing one. This is also necessary because the Chinese language is monosyllabic and without inflection (i.e., there is no need for small variations of words to express nuances in meaning as is common in European languages). For instance, English requires adding an 's' to form the plural in most cases, German adds 'e', 'er' or 'n' among others to indicate the plural. In Chinese, with few exceptions the plural is inferred by the context, not grammar. When the Japanese adopted the Chinese script, they developed a syllable alphabet (called 'Hiragana') based on reduced Chinese characters to represent the more complicated grammar found in the Japanese language. The hypothesis is that the Chinese language, being fundamentally less precise in its grammatical freedom of expression, and more constrained in its power of creating new terms for new ideas and artifacts, is less conducive as a tool to pursue science and innovation.

This is perhaps aggravated by the fact that a substantial part of one's youth is spent on memorising Chinese characters and their proper meanings. Whereas in Western languages words and their meanings can be recreated from the building blocks of the alphabet, allowing children to acquire knowledge without reference to a dictionary, Chinese children often gain access to the world only via the introduction of a new character. While some might say that Chinese minds are better trained in recognising and working with pictorial information, they also acknowledge that this practice of rote memorisation may be an obstacle to creativity.

Rote memorisation has a history in Western education, too, but this was gradually abandoned in the early twentieth century. It remains very strong in China. Apart from the linguistic context that requires memorisation, both cultural norms and the sheer number of students make exams based on reproducible facts rather than individual content and creativity necessary. Individual assessment of students is extremely time-consuming; nobody has this time in China as millions of students take tests every day. Fact-response-based elimination filters students much more efficiently, and students who have gone all the way through these exams and arrive at one of the prestigious universities are, without exception, very bright. Of course, China thinks, for the time being, it can afford to lose a few bright students in the process who were not as good at learning by heart and who, in the West, would perhaps have excelled in the arts and other creative careers.

Universities were a late contribution of Western science to China; the earliest were set up around the turn of the century in 1900. To the casual observer, universities seem to have had only a marginal effect on innovation even in Europe. This view is wrong on so many levels. Universities have produced, at least in the West, a stream of scientists, engineers, politicians, thinkers, and doers for centuries. They generate and maintain the latest science and technology even during politically or economically difficult times. They can do so only because they are, and mostly have been, independent from local worldly powers. Should a government try to impose a political agenda on academia, it renders it less effective and drives away its best people – as happened, for instance, during the Nazi Regime in Germany and the countries it invaded. This independence was itself the consequence of a legal innovation that resulted in one of the core components of the Western socio-economic fabric: corporations as legal umbrellas of collaborating individuals. The struggle between Henry IV and Pope Gregory VII in the eleventh century known as the 'Investiture Controversy' would shatter the medieval powers of kings, introduce a competing reference model of what is right and wrong, and lead ultimately to institutional reforms that allowed the development of modern mercantilism, property rights, capitalism and science.

Putting Chinese innovation into an historic context

There are two misconceptions we would like to correct with respect to innovation in China. The first is that there was quite a bit of technological progress and innovation in China. China was not static or stagnant, as is sometimes suggested. There was continuing scientific and technological progress throughout the history of China, and it was ahead of most other countries until it was, somewhat surprisingly, overtaken by Europe's explosion of modern science during the Renaissance.

The second misconception is that China was not isolated from the rest of the world and in fact had a lot of foreign influence; it just absorbed and neutralised it for the greater part of its history until Western powers were so advanced that they could no longer be ignored. In the nineteenth century, they imposed themselves upon China in what is still sorely remembered by the Chinese as a period of foreign humiliation. The infamous opium wars were waged to force China to import opium grown in India. Much of what we see in China today, politically, economically and technologically, is a response to this humiliation; it builds on an innate ability to develop its own science and technology.

The Japanese model

China is keenly aware of what is at stake. It has a close neighbor in Japan who had been in the same situation 150 years earlier but who managed to wrestle off foreign influence and become a scientifically and technologically advanced society in its own right. For many Chinese, the Japanese model is a successful escape from their own recent historical subjugation, one that ironically includes Japan as a subjugator of China in the first half of the twentieth century. While the Chinese approach is modeled after Japan, the Beijing government is also very wary of making the same mistakes.

Of course, there are also some significant differences. Japan freed itself from foreign occupation and influence much earlier than China. Japan imported foreign science and technology during the Meiji Restauration in the 1870s and 1880s, sending young Japanese students overseas and placing them into senior leadership positions upon their return - not unlike what China is doing with its own overseas students and returnees, more than one hundred years later. Japan's civil war was much shorter and more contained: Japan is ten times smaller than China, is far more isolated and controllable, and has a highly conformist culture that is even more aligned internally than China's. This made systemic change easier - a process China is still going through.

In the 1950s and 1960s, Japan had the same GDP growth rates of approximately 10% that China had in the 1990s and 2000s. Once Japan floated the Yen in 1971, its powerful MITI ministry lost control over the restructuring of the industry, and growth rates collapsed. Real estate prices were sky high in the 1980s, with even a spec of land in downtown Tokyo worth as much as entire cities in California. Here, too, some reforms in the late 1980s and early 1990s led to a crash from which Japan is still recovering. Not that Japan has stagnated scientifically or technologically: Japanese companies have actually outpaced Western companies in terms of technology development and IP production, and many consumer and business technologies are now in Japanese hands. However, for China, the prospect of seeing GDP growth decrease to 3-4% is terrifying, given it is more vulnerable and less independent than Japan at the time, and a real estate crash would destroy much of the economic wealth that China has created over the past decade, possibly leading to social unrest and hardship for large parts of its population. These are not good conditions under which to invest in scientific progress and innovation.

So, while Japan managed to avoid the fate of China in the nineteenth century, China plans to replicate Japan's success of the twentieth century. As we will see in the subsequent chapters, it is working through a long line of initiatives, policies and transformations to make this happen. As China is so different from Japan, perhaps some other countries can serve as good examples? Two of today's world leaders in science and technology, namely the US and Germany, were actually developing countries – technically speaking – in the eighteenth and nineteenth centuries. Other than the world leaders at the time – the UK and France – Germany had to import most machinery and equipment to start building its own infrastructure and industry. The US was quite liberal in using technology developed on the European continent, much to the dismay of the French and, in particular the British, who took intellectual property (IP) rights more seriously. However, by the middle of the nineteenth century, they had reformed patent law, establishing the US Patent Office in 1836. By 1900, the US had granted more than 1 million patents and had become one of the world's most advanced technological powers. Germany was considered to be the centre of science in the world.

China has a reputation for copying IP from other countries, not unlike the US in its early history, or Japan after the Second World War. Given these two precedents, copying technology doesn't seem to do much damage to future economic success. Why waste time, effort and money on innovation when technology can be apprehended for a lower cost (e.g. open innovation, if all partners agree on the legal framework) or no cost (e.g. illegal copying) elsewhere? The answer is a strategic one: if China wants to avoid getting stuck in the so-called 'middle income trap,' alongside other countries that have sub-average GDP per capita ratios and average technological and economic infrastructure, it must transform its manufacturing strength into a capability that helps China innovate right through to an advanced economy. Only a handful of countries have managed to do this; all of them were significantly smaller than China, and had significantly smaller problems. However, for China to remain a mostly underdeveloped economy with a huge past as a historic civilisation and culture is just not an option.

This was the fate of ancient Egypt, one of the top civilisations in the world until 2,500 years ago. After it was invaded and overrun, it never recovered, to a great extent because it had lost its scientific and technological superiority. This was also the fate of ancient Greece, which is the source of much Western science and philosophy, but was later integrated in the Roman Empire, and never again regained its stature as a leading power in the world. This is an unacceptable future for China.

Ancient Egypt was the centre of civilisation for 3,000 years after all, longer than even China so far. Greece was the centre of the Eastern Roman Empire until Constantinople fell to the Ottomans in 1453. The Roman Empire had also lasted 2,200 years (starting with the mythical origin of Romulus and Remus), as long as China has today (starting with the first unified China under Emperor Qin). In some ways it is easier to understand China as if it were a Roman Empire, governed from Beijing/Rome using a highly sophisticated and elaborate system of provinces and governors in the regions, relying not so much on science and innovation but rather on feats of engineering and construction. Even their symbols of power are similar: China, in Chinese Zhong Guo or 'Middle Kingdom', compares with the centre of the Roman Empire, the Mediterranean, or 'In the Middle of Land.'

Of course, Beijing wants to avoid the fate of Rome. Complacency and superior foreign invaders have contributed to its downfall. This is exactly why China is placing so much effort on innovation now, and invests in independent science and technology. China understood Schumpeter's lessons long before Schumpeter was born: agile innovators will topple the large incumbents; these innovators and their inventions must be controlled before China goes the way of the dinosaurs.

After this reminder of China's history with innovation, let us now look at the contemporary innovation scene. This is the object of the following chapter.

Chapter 3

Innovation indicators for China

The Chinese love numbers, especially big ones. It is easy to see why. For this reason, statistics concerning China must be taken with a grain of salt. This includes metrics and statistical indicators that are supposed to show the progress and success of China's science, technology and innovation. In this chapter we will try to debunk some of the statistics aimed at describing China's contemporary innovation scene.

Can numbers capture the innovation process?

For firms, to innovate is to create something novel and to commercialise it successfully in the market. This effort effectively involves the whole organisation. Considerable alchemy is involved in this complex process, which – if successful – is often heralded as a 'unique competence' of the firm. The truth is that no theory or panacea can capture – even less predict – successful innovations. Faddish slogans are crafted to seek notoriety, but they are not helpful in making the process more effective. They often fail to take into account the most important contribution to success in this field: the human factor, which is discussed in Chapter 6.

In brief, innovations fall into several broad categories:

- Non-technical innovation, such as those in the service sector (for instance the self-service model, e.g., the sharing of bicycles and electric cars in European cities, such as Paris and Budapest).
- Conceptual innovations that redefine or repurpose prior technology and resources in novel ways (e.g. the 'one country, two systems' approach to include Hong Kong, or turning a railroad station into a shopping centre, to benefit from the considerable flow of people).
- Innovation in ways of doing business and in management ('business model innovation,' such as the low cost airline easyjet, for which tickets are purchased on the Internet, or Automatic Teller Machines

(ATMs), one of the rare, truly useful innovations in the banking sector).
- Process innovation (i.e. new ways to manufacture, deliver or assemble products) resulting in reduced production costs and increased product performance and safety.
- Technology-intensive innovations, covering all aspects of R&D and new product development, with famous examples, usually covering 'firsts,' such as the integrated circuits, the Concorde, cellular phones, etc.

By far the easiest to measure are innovations that fall into the last category. Conventional indicators cover research and development (R&D) investments and the immediate results such as patents. Many of a country's innovations (including China) are non-technical in nature and therefore not captured by R&D-based indicators. Others, such as the output in patents, for example, may well overestimate the actual innovativeness of the country. In sectors in which China seems to have a strong position, such as online business and mobile telephony, these indicators are likely to underestimate the country's performance.

Non-technology-intensive innovations

In our world of radical and rapid change, non-technical innovations are more necessary than ever to complement technological breakthroughs. We need conceptual innovations to do things better and more effectively, such as providing better healthcare, particularly for older people at home (a particular prospect looming large for China), at an acceptable cost. We also need to do new things, such as effectively preserving the environment, another huge challenge in China, and producing energy renewable resources – removing the dependency on coal or other, but mostly foreign-invented, technologies. Information and communications technologies (ICT), the Internet in particular, constitute powerful tools for innovation in offerings, new activities, business models and technology dissemination. We will argue later that China, which places such high importance on innovation-led growth of new activities and jobs, is remarkably well placed to benefit from this tool, which we are only beginning to learn how to use effectively.

Non-technical, conceptual innovations include things such as the 'self-service' model, which owes nothing to technical know how. In the 1950s, the way of organising a store and its selling activity transformed how retail stores were organised and run. This could be instantly copied, facilitating its rapid diffusion. ICTs now allow firms to shift work to their customers by converting formerly in-house activities into self-service and 'do it yourself' assembly by the customer *à la* IKEA. E-banking is a great example of this shift, as are travel bookings made from home. Another historical innovation of the twentieth century has been the transformation of the state and collectively-owned economy of China into a dominant player on the world scene. China has been able to leapfrog many incremental steps of innovation that people in Western industrialised nations believed were necessary to catch up technologically.

Innovate for competitiveness

Countries and companies invest in innovation very much as an act of faith. For decades, companies and research management organisations such as the European Research Management Association (EIRMA) in Europe and the Industrial Research Institute (IRI) in the USA (an equivalent organisation does not exist in China) have had task forces study the process of technical innovation, attempting to understand which upfront conditions predict the success of the outcome, without satisfying success so far. It is nearly impossible to anticipate how successful any given investment in R&D will be. What we do know, however, is that not investing in R&D at all means giving up any chance of capturing benefits from innovation, even non-technical ones, as these are based on the exploitation of technology supported by these R&D investments.

In innovation, quality of outcome is much more important than quantity of input. The talent and motivation of the people involved in the innovation process constitute by far the most important factors of success for innovation. Within the firm, many of these factors are outside R&D, as innovation truly mobilises the whole company, including management, marketing, operations, etc. R&D makes it possible for the firm to conserve its technical know how but the generation, testing and application of new ideas and technology can happen anywhere in a firm. Innovation is, after all, a truly cross-functional effort. Therefore companies must also operate in a reasonably supportive environment: the firm's 'company culture' and the so-called 'framework conditions' must foster the innovation process. These will be discussed in Chapter 5.

All the above point to the fact that it is very difficult to truly capture the actual level of a country's innovation activity by numbers only, since so many of its components are non-technical and very fuzzy in nature. In China, two additional factors complicate matters. The first relates to China's ambition to improve its overall technological conditions quickly. The government has charted a roadmap for when certain milestones have to be met. It is therefore not surprising that China usually meets or exceeds these milestones with almost predictable regularity. We would like to believe that the Chinese Government has somehow figured out how the innately uncertain nature of technological progress and ingenuity can be managed to such exact point landings, but it is probably safer to assume that some creative tallying and tracking of criteria has been employed to allow the various responsible departments and offices to save face.

Second, it is not always clear how the many recorded invention and innovation categories are defined and measured. They may be defined differently by different parts of the government, adding to the confusion. Official records are not kept up-to-date, may disappear without comment from the public eye, or are published only in newspapers referring to 'inside sources' (e.g. the Ministry of Commerce) without ever having seen verifiable source data. The data is not always comparable from year to year, even though its presentation suggests continuity and trend, as the changes in the underlying definition are either poorly reported or not reported at all.

It is nearly impossible to debunk these difficulties looking from the outside in. They are, after all, not designed to mislead the foreign analyst, they are meant to project a growing strength and sense of innovation independence, and to reassure the people that the government is in control of technological progress. In this chapter we look at the metrics at face value as they typically describe innovative activities and offer explanatory comments in view of the particular circumstances in China. A useful basic document for foreign consumption is the OECD review of China's innovation policy, published in 2008.[1]

Metrics for innovation

The common wisdom is that 'what cannot be measured [with numbers] cannot be managed.' This is especially unfortunate in the case of innovation, since the important 'soft factors' involved are nearly impossible to measure. Many of them are based in the realm of personal judgement, grounded in professional experience and the good track records of individuals ('intuition-based management').

Still, as humans we like to reduce complexity to simple numbers, ideally into metrics that can be easily visualised. The quantitative metrics most commonly used to evaluate technology-intensive innovation include: (a) the percentage of the country's GDP invested in R&D, including the share of basic research; (2) the number of patents applied and granted; (3) the number of R&D personnel, in total and per million of inhabitants; (4) publication output; (5) number of incubators; and (6) amount of venture capital money invested. These will be discussed below. None of these are good measures for the result of innovation, but at least some of them are good proxies for the technical results of R&D.

Percentage of GDP invested in R&D

On this input measure, China shows a very strong performance. The percentage of GDP invested in R&D has rapidly increased from 0.5% in 1995 to 2.0% in 2014 (see Figure 3.1). This indicator is well on its way to achieve the country's target of 2.5% in 2020. The result-oriented political will behind this achievement contrasts with the European Union's objective of the 2007 treaty of Lisbon to invest 3% of the combined GDP of the EU countries. However, for years this figure has not budged and remains at 2%. In fact, China has overtaken Europe in terms of R&D intensity, at least as the overall average of Europe with all its member countries is concerned. China's R&D intensity is also an

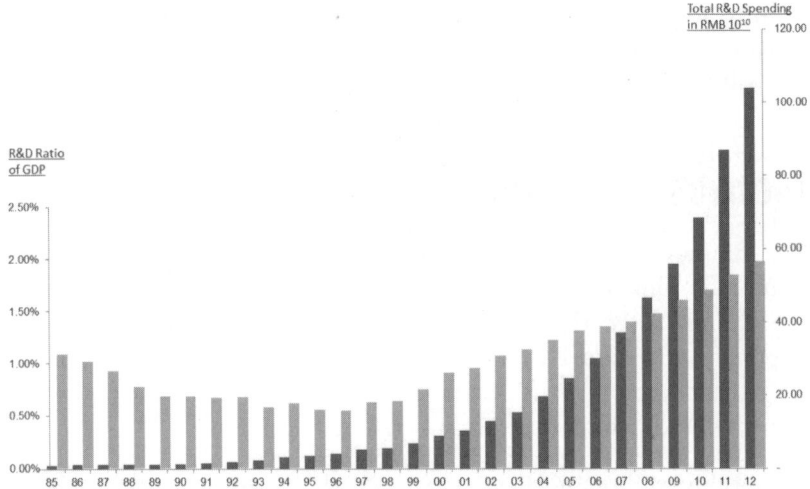

Figure 3.1 Rapid growth of R&D investments in China between 1985 and 2012, including R&D intensities, per year.

overall average including R&D-intensive areas such as Beijing, and very rural provinces that hardly do any R&D at all.

The numbers seem clear cut, but they measure inputs, about which several remarks must be made. First, out of these R&D investments, a substantial fraction (how much exactly is difficult to assess) has been used to buy equipment. Some of this sophisticated equipment, such as scanning transmission microscopes, mass spectrometers, NMR nuclear magnetic resonance, etc., has been purchased to furnish well-equipped laboratories, in order to provide arguments to persuade and attract overseas scientists and engineers to come back and work in China.

Much of this expensive equipment has a very low utilisation rate, as was the case in the USA or Europe back in the 1960s, when excessive amounts of equipment were purchased for universities and public laboratories, and the staff to operate them was also lacking. A compounding factor in China's case is corruption, buying not only large quantities of equipment but also the more expensive models, as the resulting 'commissions' are higher. As President Xi has made it his mission to fight corruption, tens of thousands of individuals are under scrutiny, but laboratory equipment is not high on his list of priorities.

China's R&D investments are heavily concentrated in a small number of areas: Beijing, Shanghai and Shaanxi. Currently, the rate of investment is slowing down in these regions to the benefit of more interior cities such as Chengdu and Xi'An, for example. This westward movement is expected to continue under the 'Go West' policy in China. The country statistics thus provide an average of a number of very different situations from region to region.

Second, throwing money at an issue is the easy part. Making an effective use of the funds, to do work leading to new activities and jobs, is a much more complex matter, requiring a broad set of conditions and a fairly long learning process. Even in the West, this requires good governance and professional judgement, sound leadership, plenty of experience and, of course, as mentioned already, competent and thoroughly skilled staff to carry out the research projects. Solid experience differentiates from just good management; research shows that people in their early forties have just the right combination of accumulated experience and tenacity to make the greatest contribution to R&D.

China, however, has still much to learn to build its know how and develop good practices for effective innovation. Many mid-career system architects and senior innovation leaders who are able to translate back and forth between customer needs and technological opportunities, are still missing. The learning is rapid but not fast enough. It still takes one year to get one year of experience. Much of this know how derives from managerial practices brought to the country by international corporations that have operated laboratories in China for more than a decade now. China's reputation for progress has been remarkable, but it has been won mostly on the easy part of R&D, following the technological lead of other countries. That will change in the future.

Output of patents

In 2012 China had a $17 billion deficit in patent-based licensing fees. Each year, it pays $18 billion to non-Chinese firms, while it collects only $1 billion. This is a clear indication that China is not a mature source of innovation. Of course, there is a lag of several years between granting a patent and a substantial amount of royalty income based on that patent. It has taken Japan well into the 2000s before it started to make money from its own licensing. A good measure of the evolution of China will be the speed with which it reduces this deficit in licensing fees.

Aside from R&D investments, a classic 'indicator of innovativeness' is the number of patents filed and granted. A granted patent is a business tool, giving its beneficiary the right to exploit the invention for 20 years. Utility models are protected in China for ten years. By and large, the number of patents per se is meaningless. Many firms have extensive patent portfolios which are unused and unexploited. What counts is the strength and business potential of the patent. It takes an experienced patent lawyer with all his judgement and probably some luck to correctly assess the business potential of a patent. For the same reason, it is very difficult to set a price on a given patent, but this must nonetheless be negotiated when a firm tries to raise money by selling its patents. As an example, in order to revise cash, Kodak sold 1,100 patents for $527 million to various companies in 2013.

Perhaps precisely because it is difficult to ascertain which patent is valuable and how much it is worth, China has been pushing hard to raise its number of patents. In 2014, it filed 928,000 invention patents. That same year, 233,000 patents were granted. Patent publications have grown at an average of 17% per year since 2005, reaching more than 2 million patent applications for all types of patents considered in 2014 – approximately three times as many as the United States (see Figure 3.2).

Mentioning these numbers without commenting on them would be misleading, to say the least. This is particularly the case for China, which has three types of patents: utility models, design patents and invention patents, the latter being the strongest form of IP protection. Utility and design patents are weak, but they represent close to three quarters of all granted patents. Next time you read about the gigantic Chinese patent application output, please remember that 75% are 'junk' or 'petty' patents. However, more recently, Chinese companies are increasingly filing invention patents as well, thus strengthening their IP portfolio and innovation capability.

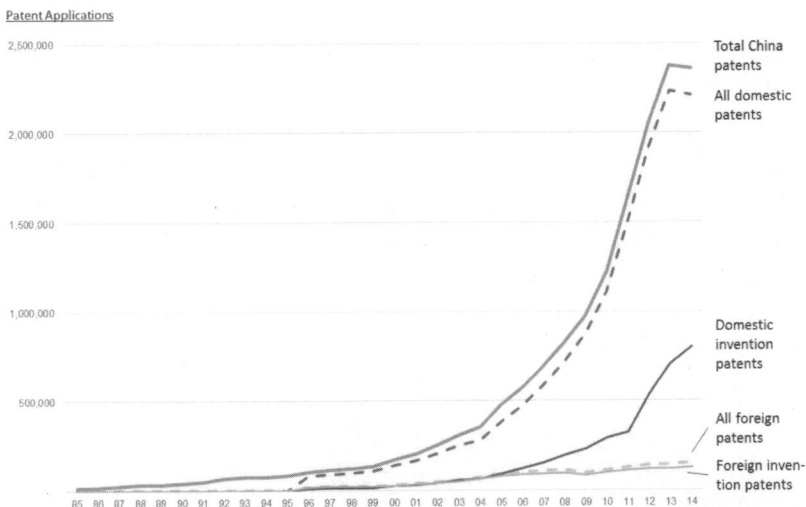

Figure 3.2 Number of patent applications in China over the past three decades.

Where do these patents come from? Geographically, the centre of gravity of patent filing is shifting from the traditional areas such as Beijing and Shanghai towards less well-known cities and regions of China. This is a consequence of the westward shift of R&D investments mentioned earlier. Guangdong Province still leads all Chinese provinces in the number of patent applications (approximately 100,000 per year), but Jiangsu and Zhejiang Province are already numbers 2 and 3; even Anhui province has more patent applications than Beijing or Shanghai. Major cities in the west, such as Chengdu and Chongqing, are growing in importance on China's patent map, which includes something like 50 inventive regions, including some 30 cities.

In terms of industrial sectors, the current share of 25% of patents in the ICT and electronics industries is expected to continue its growth. It represented 10% in 2010. Much of the new innovation activities involve internet and software, where patents are less meaningful and necessary than in certain sectors, life-sciences being the other extreme. In this sector, patents are very important for justifying the price of 'ethical drugs' until they become generics. Solid patents are also required to secure a basis for licensing to other firms. In spite of this strong role of patents, there is an innovation crisis in the life-sciences, where the number of molecules qualified by the national drug approval agencies is declining, while the investments in R&D (clinical studies in particular) steeply increase. This is another example showing that patent output does not tell the complete story.

Possibly more important than the overall evolution of the patent output is to look at the shift in patent ownership over the years. Companies, rather than public laboratories, have rapidly become a dominant force in this area and own close to 60% of patents in 2015. Among these, private entrepreneurs have increased their share, particularly in ICT, electronics, as well as for vehicles.

China's three top patent owners are ZTE, Huawei and Chery Automotive. Then follow Yulong Computer Co, in Shenzhen, Ocean King Lighting and Hong Fujin Precision, a subsidiary of Foxconn, also in Shenzhen. The quality of patents filed by these companies is well above the average of the overall output, and their commercial potential is better too. Tracking these features over time probably provides a more valid evaluation of the value and dynamics of China's patent scene.

Let's consider the many individual incentives with which patenting activity has been encouraged. For instance, monetary incentives are given to academics, granting them a bonus for every patent application. Professors with patents to their name are also more likely to get tenure. Student researchers with a patent have a better chance to gain (or retain) their hukou, or local residence permit. But not only academics can benefit: the government of Guanxi announced that it allocated $3 million to subsidise the filing of patents in the province in 2015. These subsidies can come in many forms: reduce application fees, cash rewards, expansion of legal support, etc. Corporate income tax may be reduced from 25% to 15% for companies with a high number of patent applications. Some companies offer rewards to employees who file patents. For instance, Huawei pays bonuses of up to RMB 100,000, or about $15,000 to an employee who obtains a patent.

Needless to say, these incentives work. The number of applications for the more rigorous invention patents has rapidly increased in the recent past. This illustrates one of the themes of this book: how China is able to manage top-down. After all, the goal of 2 million patent applications by 2015 was formulated back in 2010. In the West, we usually tally up all patent application numbers at the end of the year and see what the total is. In China, officials from the government and the related ministries coordinate at the start of the year to decide which province is responsible for producing how many patents in the coming twelve months, and then decides on means and ways to make this happen. No surprise then that the plan was perfectly fulfilled by 2014!

Thus, in China, the proportion of low-quality patents is high, as a result of the incentives mentioned above. Also, patents represent a notion that is relatively new to China; historically there has been no IP or even patent culture. Hence, the government also sees it as its duty to provide guidance to its people in establishing such a patenting culture. Recent trends in patent applications in growth areas such as efficiency and cost reduction, as well as environmental technologies, support this interpretation.

International patenting and PCT patents

One way to evaluate the quality of patents is to compare them using international benchmarks. This is more difficult than it seems, due to the various context-specific differences. Patent offices may be more generously staffed in one country than another (leading to a smaller backlog of patent applications), patent hurdles such as the definition of 'novelty of invention' may be in favour of easier patent granting, etc. Hence there are really just two ways to make some meaningful comparison.

One is to see how many of the local patents are filed by international companies. In China, invention patents used to be dominated by foreign companies, but this recently shifted to Chinese companies. Utility models have always been a mainstay of Chinese companies. In the US or Europe, half of all patents are applied for and granted to foreign companies, indicating how important and reliable these foreigners think it is to have their IP there.

Outward patenting may be even more revealing. In China, only 5% of domestic patents are eventually filed internationally. In comparison, more than 30% of Japanese patents are filed outside Japan. This low

Chinese percentage is an indication that most filers do not deem their contribution to be valuable enough to justify the effort and cost to file internationally. Another interpretation is that the firms filing these patents are not (yet) interested in reaching for markets outside China.

With regards to patents filed under the international patent cooperation treaty (PCT), the Western world and Japan still dominate with 92 out of the top 100 PCT filers. China had one top-100 patent filer in 2005, but added three more by 2012; the situation has been stable between then and 2015. More importantly, two were in the top ten, specifically ZTE, who took the top spot, and Huawei, who was ranked number four (Huawei was number one in 2008). This concentration in China's two competing telecoms companies suggests a continuing strong international development of that industry in the near term.

Large countries tend to have not only a large number of incoming patent registrations, but also a large number of patent filers. The total number of patents thus says little about inventiveness as measured by patent productivity per person. In a country comparison, Switzerland has the world's highest number of patent applications per thousand researchers. This is several times the Chinese number. This is due, in particular, to the high importance of high-tech industries in Switzerland.

By and large, the business potential of the large majority of Chinese patents is probably rather poor. To be fair, numerous patents elsewhere in the world are also weak (e.g. in the USA there are many 'junk patents'). Many patents should not have been granted in the first place.

Patenting Basmati rice?

In the area of illegitimate patents, there is the remarkable example of the US patent protecting Basmati rice, a staple which has been grown for centuries. The patent was granted to the US firm Rice Tec. There are several hundreds of patents on rice, mostly granted to US and Japanese companies. In 2000, appropriately, the Indian government challenged it in the courts and forced the patent to be cancelled. If this had not been the case, North American farmers would have had to pay a fee to grow a staple known for centuries.

Table 3.1 Top filers of international patents: it took Philips 2,492 PCT patents to take the number 1 spot in 2005, ZTE had to file 3,906 patents in 2012.

#	2012	2011	2010	2009	2008	2007	2006	2005
1	*ZTE*	*ZTE*	Panasonic	Panasonic	*Huawei*	Panasonic	Philips	Philips
2	Panasonic	Panasonic	*ZTE*	*Huawei*	Panasonic	Philips	Panasonic	Panasonic
3	Sharp	*Huawei*	Qualcomm	Bosch	Philips	Siemens	Siemens	Siemens
4	*Huawei*	Sharp	*Huawei*	Philips	Toyota	*Huawei*	Nokia	Nokia
5	Bosch	Bosch	Philips	Qualcomm	Bosch	Bosch	Bosch	Bosch
6	Toyota	Qualcomm	Bosch	Ericsson	Siemens	Toyota	3M	Intel
7	Qualcomm	Toyota	LG-E	LG-E	Nokia	Qualcomm	BASF	BASF
8	Siemens	LG-E	Sharp	NEC	LG-E	Microsoft	Toyota	3M
9	Philips	Philips	Ericsson	Toyota	Ericsson	Nokia	Intel	Motorola
10	Ericsson	Ericsson	NEC	Sharp	Fujitsu	Motorola	Motorola	Daimler

In brief, one must be careful not to mechanically correlate innovativeness with R&D investment and patent output. As we argue, China is on the way to being a global innovator but, by itself, China's number of patents per year is only a general indicator and not a reliable measurement of the level of innovation in the country. This indicator must thus be taken with a grain of salt. Patent activity does not properly reflect actual innovation in the country. This is due to the fact that (1) many Chinese innovations are not the object of a patent; (2) a large majority of patents (especially utility models and design patents) have a very low value; and (3) monetary incentives are redirecting inventive (or better, patenting) activity away from innovation towards short-term financial rewards.

R&D personnel

Counting engineers is also just a distant indicator for innovation, since we have seen that innovation in general – and innovation for commercial success in particular – involves much more than R&D. Innovation is a teamwork effort led by a champion. Chapter 6 looks at the crucial human factor for innovation in China.

Again, China being a very populous country, it performs well in this people-based metric, a rough indication of the country as a science and technology powerhouse. Similar to the value of patent output as an indicator, this measurement needs comment.

It is estimated that for China to reach its goal of 2.5% of GDP in R&D investments in 2020, a total number of 3.7 million researchers will be needed;[2] the current estimate is closer to 1.3 million. Chinese statistics show much higher figures. More and more R&D engineers and scientists are being employed by the private sector: The percentage of R&D personnel in the broadly defined 'business enterprise' (SOEs and private firms) has increased from 40% in 1995 to more than 60% 20 years later. The remainder is in higher education and public laboratories; this is similar to the level in Japan.

When considering the ratio of researchers per population, China has a fraction of the levels in the USA or Europe. However, China produces seven times the number of engineering graduates that the US does, so seems to be catching up fast. However, as the 2008 OECD report states, 'taking account of the quality of the human resource of this personnel in an international comparison remains a challenge'.[3] The definition of 'engineering' is ambiguous; and mechanics and technicians are often included in the headcount. The skills of some college engineering graduates are so poor that they never work in engineering but become bureaucrats or factory workers. Some statistics include two-year college graduates and part-time students; they are hardly cut out to work in R&D. That is not to say that there isn't great talent coming from some of the best schools in China; in the US, as research by Duke professor Vivek Wadhwa and others shows, Chinese immigrants have founded 12% of Silicon Valley's start ups and have contributed to 17% of America's global patents.

Only a small percentage of Chinese engineering graduates are truly prepared to operate in the conditions of Western R&D units – perhaps 10 to 20% of the graduates of top universities. This is a fairly ethnocentric view, but it is widely shared by managers of multinational firms active in China with an R&D presence (see Chapter 7). In a 2014 survey, foreign R&D managers stated that the top issue in China is no longer IP any more but the scarcity of adequate talent.

Again, at the risk of excessive generalisation, the areas for improvement in Chinese R&D personnel are as follows: First, their attitude is not very proactive: an excessive respect for hierarchy makes staff (especially junior employees) wait for bosses to give orders. Second, there is a lack of lateral communication and insufficient team spirit in conducting innovation. Third, the capacity for systems integration is poor particularly in developing large and complex structures, such as airplanes.

In brief, graduates from good universities have good content knowledge, but they lack the interpersonal skills to do well in international R&D organisations, and they may be restrained by Confucian values to take charge of innovative ideas. Young people anywhere accumulate these skills as they work, live and learn through different challenges and environments, but there is less exposure to this impetus in China. Thus the role of local management is precisely to help engineers add process skills to their content knowledge. An engineer can learn management skills (although not every engineer is destined to be a good leader!) whereas a business school graduate is unlikely to effectively learn engineering and science. This reflects China's practice of having scientifically trained persons at both the top of the Communist Party and the country.

Publication output

The number of publications in scientific journals, as well as the frequency of their citations, are additional traditional instruments to measure R&D results. International organisations collect this information and make them available in databanks, complete with statistical tools.

China calls out a separate ranking of S&T publications in a certain class of international accredited 'quality' journals (i.e. science-citation index and engineering-index journals). Between 2009 and 2013, the number of publications increased from about 26,000 to 41,000. A point of criticism is that Chinese authors target the lower-ranked among those international journals, but nevertheless an increasing number of publications now originate from China; in some scientific disciplines (within material sciences, chemistry and physics) Chinese authors are among the top contributors.

Overall, China publishes about 1.5 million scientific papers each year, with a slight year-on-year increase (see Figure 3.3). While it is reassuring to see that China is adding science manpower at a great rate, it is also somewhat disconcerting to note that over the same period, papers per million RMB have decreased from 2.34 to 1.30, and papers per capita (measured in man years) from 0.59 to 0.44.

As was already the case for patents, what counts is not the number of publications but their quality. Only quality publications introduce novelty which is then used as the basis for new research or a novel technical application. Measuring the quality of publications is at best possible to approximate via a citation count (i.e. how many times a particular paper was cited or referred to in subsequent publications). Technically speaking, this is a popularity rather than a quality measure. For what it is worth, on average Chinese papers are cited only about

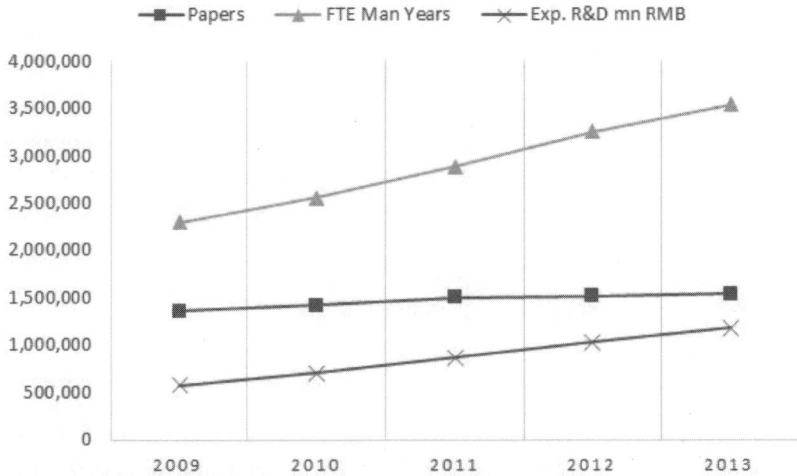

Figure 3.3 Output of publications in China, given FTE man years and R&D expenditures (in RMB million).

half as often as non-Chinese publications. In a global citation ranking, China is not even in the top 20 worldwide.

Beyond the quality of the journal or the individual papers, one way to have an indication of quality is to assess the percentage of scientific publications written between Chinese authors and co-authors elsewhere in the world, typically Europe, Japan or the USA. Work carried out in international collaboration is usually of better quality – not just with respect to China!

China's ability to copy quickly and effectively is not limited to products and patents, and extends to scientific publications as well. When Helen Zhang, a Zhejiang University-based editor of a journal, introduced CrossCheck, a text analysis software to spot plagiarism, she was astonished to find that 31% of the papers in her journal had used unreasonable copying and plagiarising. In some disciplines, such as computer science and life sciences, this increased to 40%. Not only are academic papers important for obtaining tenure, they often also come with a cash bonus paid by the university department, so aspiring professors focus on getting published often and wherever they can.

Some professors make it mandatory for their graduate students to publish with them as co-authors before they are eligible for graduation. Conflating or combining different parts and pieces from various papers and merging them into a new paper is not uncommon. The peer-review system (which was invented in the West, one should say) does a poor job of detecting such violations, and is being stretched with the explosion of new journals and online publications.

There is research that correlates the output of scientific publications with economic wealth, but the causality is less than clear. As in the case of the individual innovation process, how exactly scientific ideas transcend into innovation (which, at least here is some general agreement, cause economic growth and thus wealth) is still a matter of much debate and confusion. Furthermore, many business innovations come from online applications which are never published, especially not in scientific journals. Here, again, we measure something that is at best remotely associated with innovative activities, and is relevant mostly only to technology-intensive innovation.

Science parks and incubators

There are 115 university science parks and over 1,600 technology business incubators supporting start-up technology companies in China. Over 80,000 enterprises are housed in these institutions, generating 1.7 million jobs. This seems an impressive statistic, which evokes a most vibrant innovation scene. Again, an inside look often reveals more about the quality of the action.

Many incubators are essentially office parks for rent without much focus on the entrepreneurial nature of the firms or actual start-up specific management advice. The same is true for China, masked behind the busy-ness of Chinese everyday life. Incubators are often multi-story buildings in an urban environment. In the lobby of the building, the visitors are taken to see the displays on the activities or products of the start ups in the incubator. They are then taken to a conference room with a video presentation, describing the activities of the incubator. In many cases, however, when walking around these 'incubators,' one quickly realises that they lack activity and that they essentially shelter deserted offices. Rarely does one see the rumbustious and feverish activity usually associated with entrepreneurial teams working to launch their business. These buildings seem to be very much the result of real estate speculation and corrupt practices, to enrich individuals or institutions and participate in the considerable 'hype' currently attached to the concept.

Furthermore, there is frequently a *'mélange des genres,'* as defined by Western standards. For example, a university professor, involved in a start up, has part of his university laboratories function as the R&D department of the young firm, with the staff, equipment and space paid for by Chinese taxpayers. Such practices are not perceived to be suspect; they are part of the normal way of doing business, whereas, in the West, this is considered to constitute a conflict of interest and suspect in that people are not working on a 'level playing field.'

A minority of the institutions are *bona fide* incubators doing useful work to house and help productive entrepreneurial teams, in order to develop their activities. Examples of those include Tianfu Software Park, in Chengdu, People Squared, in Shanghai and Hxlr8r, in Shenzhen, for devices and hardware, the favourite region for the 'makers' movement.' The latter will be discussed in Chapter 8. These truly help young firms to launch and develop. They have a professional staff working to support this mission.

For instance, Tianfu Park was launched in 2005 and houses more than 1,100 software enterprises. Each year, more than fifty firms come to benefit from the environment and services of the park. Beyond providing an infrastructure and stimulating environment, the park actively participates in incubating businesses through advisory work, training, workshops, as well as helping with administrative red tape, such as company registration and relations with the government. One of the recent success stories of the park is the tap4fun electronic games company. This success is encouraging others in the gaming sector, in which China is extremely active.

Investment in long-term (basic) research

As defined by the Frascati Manual,[3] basic research consists of experimental and theoretical work aimed at creating knowledge about fundamental principles and phenomena. China is currently investing small amount of funds in long-term scientific research. Reported numbers vary from as little as 1% of total R&D investment to about 7%, compared with 15% in the case of the USA. Such long-term research is typically carried out in the institutes of the Chinese Academy of Sciences or Key National Laboratories. There are few independent observers of what exactly is happening in this national research centre. Some cynics have commented that rather than 'research and create,' as it is the mission for public Western laboratories, Chinese labs pursued a mission of 'search and recreate,' implying not much innovative work is taking place.

In a 2010 editorial in the reputable magazine *Science,* two prominent Chinese scientists, Rao Yi of Peking University and Shi Yigong of Tsinghua University, wrote that 'to obtain major grants in China, it is an open secret that doing good research is not as important as schmoozing with powerful bureaucrats and their favourite experts.' No wonder then that 'a significant proportion of researchers' are more often to be found building and entertaining these connections than doing the actual science itself.

Many Chinese academics and government 'think tanks' argue that China must invest much more in basic research. Such research is essentially curiosity-driven, although it may eventually lead to patent applications. On the one hand, one could say that basic research is the cheapest investment a country can make. On the other hand, research has always been considered an opportunity cost (i.e. money that could have been spent with quicker returns elsewhere). It will be interesting to watch what China will do in this area. Research results are a highly mobile good, so basic research carried out in one country can be put to use fairly easily in China, essentially concentrating on assimilating research results, generated outside the country, in order to apply them in market-oriented applications. But this *free-rider behaviour* destroys goodwill between countries and scientific communities.

The topic of long-term research may be the object of geopolitical pressure. One example happened with Japan, in the 1980s. The then prime-minister Yasuhiro Nakasone was asked to shoulder more of the world's scientific research. Japan's response, which was discussed at the G7 conference in 1987, was to create Human Frontiers, an agency founded in 1989 and headquartered in Strasbourg, France, to provide research funding to intercontinental research teams in the area of life sciences. Perhaps China will follow a similar path.

The venture capital industry

A vibrant venture capital (VC) industry is often considered a good indicator of innovativeness of the country. On one side of the spectrum is highly innovative Israel, which has a VC industry comparable with that of Great Britain, a much larger country. On the other hand, Switzerland, which does not have much of a VC industry and does not even recognise the status of limited partnerships, common to VC funds, is regularly ranked as a top innovating country.

The VC industry, invented in 1946 by the French American General Georges Doriot with his American Research Development, is not very profitable overall. In this numbers game, it occasionally has a great success, with a lucrative IPO, so that the media sees the industry as being shrouded with a bit of magic. In fact, even in the USA, less than 10% of start-up firms receive 'smart money' from a VC to finance their development.

The VC industry in China has more than 2,000 funds. It is said that a start up has an easier time finding funding in Beijing or Shanghai than in Cambridge or Silicon Valley. Easy money, however, is not smart money; it is certainly not guaranteed to be well-invested money.

In 2001, the government started to enact measures to regulate the VC industry. The majority of the money comes from public sources; the government calls for proposals from fund managers to administer a fund. In 2015, the rules changed again to allow insurance companies to invest in VC and a fresh $6.5 billion was injected into VC funds by the state. Together with other financial aspects of innovation, the VC industry in China is further discussed as part of the framework conditions in Chapter 5.

In brief, conventional indicators considering R&D investments only take into account technology-intensive innovation. Many of China's innovations are not technical in nature and therefore are not captured by the above indicators for R&D activity. Following the above discussion of the metrics classically used for evaluating innovativeness in China, these indicators should not be considered at face value without appropriate caveats. Other indicators, such as the patent output, may well strongly overestimate the actual innovativeness of the country.

In sectors in which China leads the way, for instance the Internet and mobile telephony, these indicators are likely to underestimate the country's performance in this area. As China innovates in the ways of practising innovation, it offers useful lessons for the non-Chinese world to learn. It is up to the world to seize that opportunity. In all, conventional innovation metrics, which are essentially geared to technical innovations, must therefore be used and interpreted with particular care.

---

[1] 'China Review of Innovation Policy' (OECD, Paris, 2008).

[2] *China's Three Waves of Innovation*, Report by Koros Future, 2013.

[3] Frascati Manual (OECD, 2002).

Chapter 4

Government and firms as key actors of the innovation scene

The main actors of China's wealth-creation process are companies, whose actions are framed and regulated by the government and its bureaucracy. The interaction between companies and the public sector is probably unique in the world. Companies, even those that are private, are under some sort of public sector guidance. This chapter describes governmental institutions and companies, as well as the scope of their general action, setting the general context for innovation. In Chapters 8 and 9, we will revisit them to discuss their policies and practices specific to fostering and practising innovation.

The government apparatus

The Chinese government is highly centralised. It also has a very broad span of control. For example, state-owned banks are led by senior officials of the Communist party. When it comes to the 'policies' and 'framework conditions' for economic activity, Chinese bureaucracy has a very strong role in shaping economic activity and conducting the China orchestra. This is a legacy from the system of the former USSR (e.g. Gosplan) when it played the role of model and teacher for the then recently founded People's Republic of China. In the past few years, this legacy has somewhat decreased and the influence of private enterprise has risen. The current leadership is stressing entrepreneurship, but it remains to be seen whether this will constitute a strong pattern in the near future.

The government system has been in place since the proclamation of independence in 1949. Since then, China has been ruled by the Chinese Communist Party (CCP), which today counts close to 90 million members.

Administratively, China is divided into 23 provinces, which present much diversity, five autonomous regions and four municipalities. China's current constitution dates from 1982.

The leadership

At the top of the country is the President of the People's Republic of China. Currently the post is held by Mr Xi (first name Jinping), who was born in 1953; he trained as a chemical engineer at Tsinghua University.

Mr. Xi made his political career in the provinces of Fujian, Zhejiang and Shanghai. Since 2012, Xi Jinping has been General Secretary of the Communist party and Chairman of the Military Commission, as well as, *ex officio*, member of the Politburo Standing Committee, which is China's top decision-making body. He succeeded Hu Jintao.

President Xi was 15 when his father was jailed during the Cultural Revolution in 1968. He is sometimes called a 'princeling', i.e. his parents belonged to the first generation of the revolutionary elite.

Xi's first trip as president outside Beijing was to Guangdong, where his speeches focused on strengthened military and economic reforms, paying homage to Deng Xiaoping. Generally, however, his policies step up censorship and control – the opposite of Mr Gorbachev's 'glasnost' and 'perestroika' – while his vigorous anti-corruption campaign has, so far, often targeted his rivals, including several of the political supporters of Jiang Zemin. This campaign seems to be gaining momentum, as several thousand party officials have been the subject of enquiries as of 2015.[1]

President Xi has articulated the slogan of the 'Chinese dream' to indicate the will for China to regain the eminent place it had in the world until the nineteenth century. In May 2014, he strongly encouraged Chinese firms to carry out three transformations: (1) move from 'made in China' to 'created in China'; (2) move from speed to quality and (3) move from product to brand. These three imperatives are increasingly in the background of policies that are implemented by the sometimes conflicting interconnected web of China's public actors.

The second most powerful person in China is Prime Minister Li Keqiang. Born in the rural province of Anhui, he had to spend years on a farm during the Cultural Revolution. In 1978, he was accepted to study law at Peking University, which had just re-opened. Twenty years later, he became China's youngest governor, in Henan. During his tenure, the scandal emerged that thousands of persons had contracted HIV from contaminated blood from a government blood donor campaign. Considered as a protégé of former president Hu Jintao, he was his likely successor until it became clear in 2007 that Xi Jinping would get the job. On economic reforms, Mr Li is very much in charge. For example, he was the champion for the Shanghai Pilot zone for trade, created in 2013 (see Chapter 8).

The institutional system in Beijing

The highest ranking body for coordinating science and technology (S&T) is the State Council. Founded in 1998, it is chaired by the Prime Minister. It is primarily responsible for the elaboration of China's S&T strategic plan until 2020.

The key elements of the administration and policy-making apparatus are:

The Ministry for Science & Technology (MOST). The Ministry of Science and Technology (MOST), formerly State Science and Technology Commission, is the ministry that coordinates science and technology activities in China. Furthermore, it supports basic research and development (R&D) and administers national science & technology (S&T) grant funding programmes. It has its own newspaper, *Science and Technology Daily*. At the local level, MOST offices are called 'science and technology commissions' (STCs).

The National Development and Reform Commission (NDRC). The NDRC is a macroeconomic management agency under the Chinese State Council with broad administrative and planning control over the Chinese economy. Its functions are to study and formulate policies for economic and social development, maintain the balance of economic development, and to guide restructuring of China's economic system. The NDRC leads central government coordination of new industries such as SEI (the Strategic Emerging Industries, see Chapter 8) interagency work and guides overall policy development. It has a specific department for high technology industries. At a local (province) level, NDRC offices are called 'development and reform commissions' (DRCs).

The Chinese Academy of Sciences (CAS) reports directly to the State Council. It has a strong influence on S&T policies and has responsibility for the commercialisation of R&D results.

The Ministry of Education (MOE) has comprehensive responsibility for higher education and university research.

The Ministry of Industry and Information Technology (MIIT) was established in 2008 to assist in the development and regulation of internet, wireless, broadcasting, the postal service, communications, production of electronic and information goods, software industry and the promotion of the national knowledge economy. The MIIT combines the old Ministry of Information with the former Commission of Science, Technology and Industry for National Defence, and the State Council Informatisation Office. At a local level, MIIT offices are commonly referred to as 'economic and information technology commissions' (EITCs).

Ministry of Commerce (MOFCOM), formerly called the Ministry of Foreign Trade and Economic Co-operation (MOFTEC), is an executive agency of the State Council of China. It is responsible for formulating policy on foreign trade, export and import regulations, foreign direct investments, consumer protection, market competition and negotiating bilateral and multilateral trade agreements. It coordinates with other agencies to support industrial policy development and implementation. At the local level, MOFCOM offices are often referred to as 'commerce commissions' (CCs) or 'departments of commerce'.

The Ministry of Finance (MOF) is the national executive agency of the Chinese central government that administers macroeconomic policies and the national annual budget. It also handles fiscal policy, economic regulations and government expenditure for the state. It serves as the

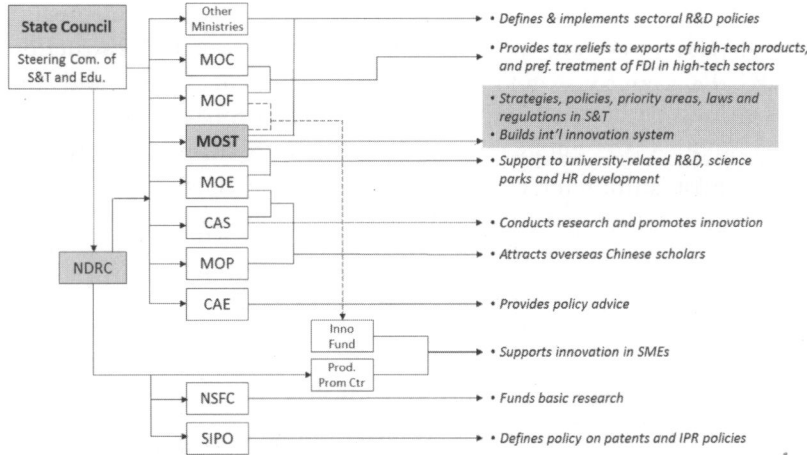

Figure 4.1 The interplay of key ministries and agencies in China's innovation system.

Source: OECD Reviews of innovation Policy: China, 2007.

primary agency managing available funds for development, allocating funds for public R&D activities and infrastructure. At the local level, MOF offices are referred to as 'departments of finance'.

The State Administration of Taxation defines the fiscal incentives.

The State Intellectual Property Office (SIPO), also known as the Chinese Patent Office, was founded in 1980 and is responsible 'for patent work and comprehensive coordination of the foreign related affairs in the field of intellectual property' (SIPO website).

The State-owned Assets Supervision and Administration Commission (SASAC) promotes the governance of state-owned enterprises (SOEs) and oversees more than one hundred such firms, e.g. the China National Nuclear Corporation, China Aerospace Science & Technology Corp., the Aviation Industry Corp., and the China National Petroleum Corp.

The Banking Regulatory Commission provides the framework conditions under which takes place the financing of innovation, in particular Venture Capital (VC).

The banking system is state controlled. The largest state-owned commercial banks are the Bank of China, the China Construction Bank, the Industrial and Commercial Bank of China and the Agricultural Bank of China. The executive branch of the government occasionally orders banks to grant low-interest loans to companies to finance acquisitions and their expansion outside China.

The State Administration for Industry and Commerce (SAIC) is in charge of supervising and regulating the market, including enforcing the corresponding laws and anti-monopoly rules.

In addition to the public institutes, discussed in the following section, other national, public actors include the following:

- *The National People Congress* (NPC) is China's Parliament, which counts close to 3,000 members, mostly belonging to the Communist Party. These are selected for a term of five years; they sit each year for one session.
- *State government bureaucracies of the provinces* often compete with each other, and occasionally are in opposition with Beijing.
- *The People Liberation Army* has 2–3 million people and a budget representing 1.4% of China's GDP. This is the world's second largest military budget after that of the USA; it represents a sum estimated between $100 and $160 billion (roughly a third of that of the USA). Some think that the actual budget is substantially higher. In 2015, China has confirmed that it is building a second aircraft carrier. It is the world's fifth largest exporter of weapons after the USA, Russia, Germany and France.

The defence establishment has an array of laboratories, including the State Administration for Science, Technology and Industry for National Defence (SASTING). Some of these laboratories were moved away to Sichuan province, from the Harbin region, when relations with the USSR soured. Not much information is available on their activities.

Public institutes

China has 104 institutes run by the China Academy of Sciences (CAS), founded as early as 1949. They have roughly 60,000 employees, of which

43,000 are research professionals. In a recent reform it was announced that these institutes will be put into four classes: basic research, applied research and commercialisation, big science centres with large facilities and institutes working on issues specific to a region. The institutes will either slot into one of these categories, or be disbanded.

The activities of the institutes are fairly evenly distributed among mathematics and physics, life sciences, technological and engineering, chemistry, earth sciences and information technology. The laboratories of the CAS have been encouraged to contribute to the country's economic growth. The computer manufacturer Legend came from one of the CAS Institutes. It was incorporated as Legend in Hong Kong in 1984, later changed its name to Lenovo and bought the personal computers divisions of IBM, including the Thinkpad. It is now the world's largest producer of personal computers operating in more than 60 countries.

Overall, the public research institutes roughly represent 500,000 people. Their activities have a strong focus on farming/environmental issues. Their output in patents and publications has increased in recent years, but, as mentioned in Chapter 3, their quality is often questionable. Their share of the total Chinese patent output has, however, decreased from 8% in 1995 to 4% in 2014, underlining the transition from state-run innovation to enterprise-led innovation.

The role of the institutes is important in developing a population of experienced scientific researchers who benefit from the various technology R&D programmes as govern from Beijing. Many of them are being transformed into enterprises charged to transfer technology to commercial applications, an activity which will continue to be a main focus in the coming years.

Example of recent policies and reforms

The fact that China joined the World Trade Organization (WTO) in 2001 was a decisive step in the recent history of the country. It allowed China to fully participate in world trade and to greatly enjoy its resulting benefits, while forcing the country to align its laws and start moving its practices to be compatible with the prevailing Western way of doing things. One of the legal aspects concerns intellectual property. In this regard, the laws became world standard, but it took the courts a fairly long time to practice them in an effective way. Note that China joined the World Intellectual Property Organization (WIPO), based in Geneva, much earlier, in 1980, adopting the Paris Convention on IPR in 1985. The intellectual property system will be discussed in the framework conditions (see Chapter 6).

Anti-trust law

The Anti-Monopoly Law (AML) concerns firms in their competitive environment. The AML was enacted in 2008, drawing on many elements of anti-trust laws from Europe, USA and Japan. In many ways, however, it lacks legal clarity. Even more crucially, it excludes monopolies sanctioned by the state, as well as complete sectors dominated by SOEs. This therefore gives ample scope for courts and government agencies to use the AML as a tool to protect national companies and help the domestic base of Chinese champions, especially as they become more active outside China.

One of the controversial aspects of the AML is that the current text can be used to force a non-Chinese corporation to license a specific technology to an emerging Chinese competitor on the ground that foreign IP is preventing the legitimate development of the latter. In 2015, foreign carmakers have been targeted by the Chinese authorities for alleged non-competitive behaviour and practices. This could be understood as a signal that the government is not satisfied with the state of the market.

Oversight of mergers and acquisitions

The SASAC (State-owned Asset Supervision and Administration Commission), mentioned earlier in this chapter, examines investments and purchases by non-Chinese firms. In 2006 and the following years, laws were put in place to reinforce the approval process. For example, in 2007 Alstom was blocked in its attempt to buy Wuhan Boiler. Seven sectors are considered particularly sensitive: defence, electricity production and distribution, oil, coal, telecommunications, civilian aircraft and sea transport.

Reform of the hukou system

The broad plank of reforms, particularly in the economic and financial areas, announced at the November 2013 Third Plenum, has not delivered many significant policy changes so far. One planned reform concerns liberalising the *hukou* system, a kind of an internal passport for people migrating within China to find work, usually in the cities. The government thus has a way to control the flow of people within the country. This document allows access to social services, such as hospitals and schools, but at rates higher than that paid by locals. There is a movement in China claiming that the hukou system is against the constitution, as it discriminates among citizens. A possible way to solve the problem may be that the hukou requirement will be progressively lifted, first in the medium-sized cities, then later in the larger ones. In this case, the municipalities will have to find the additional funds to provide social services to the migrants.

Environmental law

In the area of environmental protection, much is to do in China. The serious air pollution of Beijing is only one manifestation of the ecological crisis in the country, as water and soil pollution, less visible, are extremely serious. The country has sacrificed its environment to industrialise at a relentless rate and the cost of clean up will be huge; in the meantime, serious crises will emerge. China's top leadership is keenly aware of this. In 2014, the Prime Minister declared 'We need to wage a war on pollution'. In April of the same year China overhauled and updated its legislation for environmental protection, dating back to 1989. The revised Environment Protection Law (EPL), which came into force on 1 January 2015, provides for higher standards and stiffer fines, as well as much more thorough disclosures. The real effectiveness of the new law will, however, come from the way it will be implemented and enforced.

Chapter 5 of this new law grants a vastly enhanced role to civil society. In effect, it transfers some of the responsibility of implementation to not-for-profit organisations, which have rapidly grown in numbers and sophistication in recent years.

Cybersecurity

A pattern in China is a strong emphasis on cybersecurity. A Central Cybersecurity and Information Leading Group was put in place in 2014. It is headed by President Xi and includes 21 ministers from relevant ministries and commissions. It focuses on long-term ICT security and industry strategy and planning, as well as inter-agency coordination.

Europe and the USA are concerned that, if fully implemented, certain of the resulting policies would prevent Western firms from participating in the IT equipment for the banking sector, including Automatic Teller Machines (ATMs), point of sales terminals and smart card readers.

Anti-terrorist law

Consistent with a general stiffening of domestic policies on internet control and censorship, in Spring 2015 the National People Congress voted China's first anti-terrorist law. This takes a very extensive view of terrorist activities and puts in place anti-terrorist structures with sweeping discretionary powers. This is primarily aimed at the 10 million Uyghurs of the autonomous region of Xinjiang, on the eastern border of China, which became part of the Empire in the eighteenth century. Such a law, it is feared, will help legitimise violent repression. In 2013–2014, there were three times more Uyghur victims than those of Han origin (i.e., the main Chinese ethnicity).

Local government

Regional governments play an important role in setting up and encouraging innovation. For example, each province mirrors the central government's ministries, including a Science & Technology Commission. Such a commission is also found in municipalities and certain of them, such as those of Shanghai, are particularly powerful and effective. For example, the latter is investing substantial amounts of resources in areas such as the health sector, including biotech, and additive manufacturing ('3D printing'). In this area, the intent is to have in Shanghai 100 'hackerspaces' for the 'makers movement' described in Chapter 9. Overall, it is estimated that 60% of the investment in R&D/innovation is made by central government, with the remaining 40% coming from local government.

Initiatives are also being taken at the local level. As an example, in 2014, the Beijing Government adopted measures to improve the transfer of technology from the city's universities to firms. The Beijing region has more than 85 universities, of which 34 are directly affiliated to the municipal government. Decisions in this matter have thus been transferred from the government to the universities. There are also dramatically increased incentives for researchers to commercialise their research results. This includes the possibility for researchers, after approval by their university, to be seconded to work in start ups. The duration of this 'entrepreneurial leave' is open-ended and its termination is left to the university.

The same reform confirms the appointment of a 'technology manager' in each university. This is easy to do; the difficulty is to have competent and motivated staff with the appropriate background and training to take on the complex job of effectively moving research results to the market. In the West, it is difficult to staff technology transfer units with people who have a good understanding of the processes at hand and of the business, the technical aspects and personnel issues. Japan, which created Technology Transfer Offices (TTO) several years ago, faced a similar scarcity of appropriate personnel. Effective management development and well-prepared 'traineeships' may mitigate this situation.

Huaxi: a village of share-holders

Huaxi is located in Jiangsu province, 80 km from Shanghai. It offers an example of what a local government can do.

The village founded a multi-industry company that the local Communist party chief Wu Renbao took to the stock exchange in 1998. The 2,000 original villagers became shareholders of this company controlling businesses worth $6 billion dollars ranging from steel to cigarette papers. The parent company, Huaxi Holding, is incorporated in the Cayman Islands. The village is often termed the richest village in China, with villagers having access to apartments and cars, a long way from their origins as a farming community.

In this shareholding feudalism, if the registered villagers leave the village, they lose everything (the minimum villager's net worth is about $100,000). Also the non-registered persons (some 35,000) who work there do not share in the monetary advantages and earn a normal Chinese salary. Apparently villagers are not allowed to speak to foreigners or the press.

As an attraction, a 72 storey/328 metre skyscraper was built in the village. This is to help the transition to becoming a socialist tourist 'hot spot', at a time when the future of the steel plant became uncertain.

The university system

Universities are administered and financed by the Ministry of Education (MoE). The local governments also have some control. In 2015, it is estimated that there are more than 11 million students enrolled in some 1,500 colleges and universities in China. In the Beijing area alone there are 83 institutions of higher education. Of these, 34 are affiliated to the central government, while 49 report to the municipal government.

Nearly 7 million students obtain a degree from China's university system every year. Thirty one per cent of the students have a degree in engineering, compared with 5% in the USA. By 2030, China anticipates having roughly 200 million college graduates.

China's government is encouraging mergers between universities, as well as creating new ones. The 211 programme calls for the creation of 100 new universities in the course of the twenty-first century. In this process, often no more than 18 months separate the start of the construction of the campus and the beginning of classes. This frenzy to build is common in China. The creation of new universities, however, will hit a bottleneck, as it takes much longer to develop effective professors than to build new campuses. As a result, it will take a lot of time and effort for China's new universities to reach a high standard. The national programme '985' aims to create a small number of world class universities in China.

Going back in time, the difficulty in developing faculty stems in part from the devastating effects of the Cultural Revolution on higher education from 1967–1976. During that time, the number of students in the entire country fell below 50,000. Universities were closed for six years. It was only in 1977 that the Gao Kao (University Entrance Examination) was re-instituted by Deng Xioaping. In 1978, 400,000 students entered university. It took until the mid-1980s until a new cadre of faculty was in place and the numbers of students had increased significantly. The impact of that period is still being felt today.

China has one of the world's best-known ranking exercises for universities, known as the 'Shanghai ranking'. Since 2003, this has been carried out by the Shanghai-based Jiao Tong University. According to this ranking system, the first-ranked non-US university is the University of Oxford in Great Britain, and the first in Europe is ETH Zurich. Peking University is the first mainland Chinese university, ranked in the 100–150 bracket, closely followed by Tsinghua and Jiao Tong University itself.

This ranking is heavily based on awards, such as Nobel prizes as well as Field medals for mathematics. As with any ranking system, it lends itself to criticism concerning its 'methodology' and the criteria selected.

Each year in June, close to 10 million young people take the Gao Kao entrance examination into the university system. However, many students forgo a Chinese university education and also apply for overseas college slots. That so many students go abroad for higher education is an indictment of this system. Today, almost one in three international students in US universities comes from China. There were close to 300,000 in 2014. One of them is the daughter of President Xi, who enrolled at Harvard University as a freshman in 2010.

Chinese universities have embraced massive open online courses (MOOCs). Peking University and Tsinghua are leading the way in this area. Some of these courses are available on coursera and edX. It is interesting to speculate as to what effect this may have on the Chinese university system.

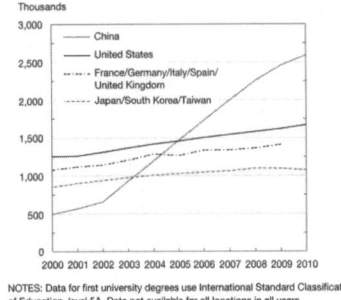

	USA	CHINA
Gov't spending on education	5.62%	1.91%
Tertiary enrollment	72.6%	7.5%
Spending per college student	22.04	90.07
Tertiary students abroad	57,000	645,000
... per thousand people	0.183	0.48
Journal articles	211,233	29,186
Top-50 university per 100 mn pop.	6.67	0.0767

NOTES: Data for first university degrees use International Standard Classification of Education, level 5A. Data not available for all locations in all years.

Figure 4.2 Comparing China and the USA's statistics on higher education.
Sources: left figure by National Science & Engineering Indicators, 2014; right table by NationMaster, 2015: China and USA Compared (using most recent available data).

Several universities are putting innovation and entrepreneurship at the centre of their curriculum and educational activities. Another way for universities to introduce more of an innovative and entrepreneurial spirit into their operations is to launch or enlarge programmes on design and creation and develop 'maker laboratories'. As will be seen in Chapter 9, 'makers' are students who tinker with devices, usually in telecoms, computers and software, assembling novel gadgets, including using additive printing. Universities such as Tsinghua and Tongji participate in that movement, as well as the Shenzen accelerators and innovation laboratories. Universities set up laboratories for makers to experiment. The maker movement received a great notoriety boost from China's Prime Minister when he visited a makers place in February 2015. The top leadership sees the positive side of bottom-up innovation, very much in line with their encouragement of entrepreneurship.

Chinese universities: can do better

Universities are all about excellence in teaching and research. An additional role is technology/knowledge transfer. China's universities need to develop excellence in research. PhD students must develop a passion for their discipline and research topics. Currently, a student selects a PhD advisor more for his/her notoriety and influence than for the research. This may be a consequence of Confucianism.

The Chinese have a real fascination with US universities. Universities usually pay better salaries to faculties who have obtained a foreign doctorate than to those with domestic PhDs. The challenge is that developing effective higher education institutions takes longer than building a campus. The quality of governance and faculty and a research 'culture' are key. These take time.

On the issue of freedom of expression in universities, it will be worth watching what will happen in coming years to the recent trend of clamping down on 'Western values' by the Minister of Education. Late in 2014 President Xi called for more 'ideological guidance in schools and universities'. Universities have been reminded to strongly encourage 'patriotism and love of the party', as well as being careful in the way they handle a number of topics in the classroom. Foreign textbooks are to be abandoned in favour of Chinese domestic textbooks.

This is consistent with the recent reinforcement of the 'Great Firewall' mentioned elsewhere. In 2014, Reporters without Borders ranked China 175th out of 180 countries.

Figure 4.3 Rankings of Chinese Universities.

Judiciary system

The intellectual property (IP) system on framework conditions is discussed in Chapter 5. The purpose here is to present the overall legal infrastructure.

The Supreme People's Court is the highest judicial organ of the state. It has four primary functions: interpretation of law, adjudication, legislation and administration of the judiciary. The Higher People's Courts hear first instance and appellate cases involving civil, administrative and criminal law. Each province, directly administered city, and autonomous region has such a court. In commercial matters, a higher court hears first instance cases where a large amount of money is in dispute and bankruptcy cases where the company involved received its business license from the provincial level State Administration for Industry and Commerce.

Then there are the Intermediate People's Courts. These are located in a municipality and have original and appellate jurisdiction in criminal, civil, economic and administrative cases.

Finally, there are the Basic People's Courts in rural areas or municipal districts. They have original jurisdiction in certain civil, economic, administrative and criminal cases.

For many years the Chinese Communist party did not like private companies. The relations between these two actors have very much improved over time. After discussing aspects of the government let us turn to the firms, which are now perceived by the government as an essential tool to provide new activity and jobs.

Firms' managers must engage with the authorities

A company had built a chemical plant in 1995 near Shanghai. The plant presented some nuisance for the neighbourhood. Gradually, housing developments were built in the area close to the plant, so the local authorities decided that the plant had to move. They could have abruptly notified the company that operations should stop within a month or two. Fortunately, the plant managers had maintained a good dialogue with the local authorities, inviting them to visit the plant, and occasionally paying them a visit. This tactic paid off handsomely: it allowed the two parties to discuss and plan the move, obtain new land at a reasonable price and carry out a smooth relocation.

Companies

Compared to many countries, China's gross domestic product (GDP) has a relatively low percentage in the service sector (45%), while industry accounts for 45% and agriculture for 10%. In the evolution of the country towards more services and more value-added offerings, China's companies have an eminent role to play and numerous opportunities to seize.

Like most countries, China accounts for private firms, large and small, including start ups, as well as state-owned enterprises (SOEs). A difference is that, in China, private companies are still heavily influenced by the public sector. Another segment is constituted by the large, non–Chinese companies operating in China. These represent one-third of China's GDP; the lion's share of its exports are so-called 'high technology' products. These are discussed in Chapter 7.

Between 1949 and 1978, all companies in China were either state-owned or collectively owned. They were founded during the era of Mao Zedong. Employees just followed orders. All managerial decisions on what to produce, at what price, were dictated by the government. These firms had no marketing or R&D departments.

The economic reforms initiated by Deng Xiaoping in 1978 introduced dramatic changes. Management teams acquired some autonomy and began to be rewarded on the basis of the results of their firm. Private companies could be created alongside public ones. In 1992, more than 100,000 government officials resigned from their jobs to go into the commercial world. The first joint venture between a Chinese firm and a non-Chinese company was concluded in 1979 with the Swiss company Schindler, producing elevators and escalators, headquartered near Lucerne. By the year 1983, close to 1,000 joint ventures had been formed, with investors mostly from Hong Kong, Macau and Taiwan.

During the period 1987 to 1992, numerous SOEs became equity companies. Until then this had been restricted to specific provinces. In 1992, FDI was allowed throughout the whole country and represented close to US $500 million in 1996.

In the ten years following 2004, the number of SOEs (2 million) halved, while those of private companies rose from 0.5 million to close to 4 million. Thus, the emergence and growth of small and medium-sized enterprises (SMEs) are relatively recent.

Provinces have adopted different attitudes towards companies. For example, the Shanghai region has chosen to attract large foreign multinational companies. Driving for hours along the Suzhou industrial park, one sees an industrial plant with the logos of corporations from all over the world. A different approach was adopted by Wenzhou, a city which is part of a metropolitan area of close to 8 million people, in the south of the Zheijang Province. In 1979, for the first time in China, a licence was granted for a private retail business. The city was allowed to set up individual and private enterprises, as well as shareholder cooperatives. In the 1990s, this municipality elected to encourage people to start their own companies. Today, 10,000 small firms exist in that region, organised in a way similar to the 'districts' of northern Italy, and often using the Internet as a channel for distribution and sale. Today, 95% of the local economy is in the private sector. Wenzhou is generally considered to be the birthplace of China's private economy.

Examples of firms

The large and dynamic economy has seen the creation and growth of very diverse firms, a sample of which are briefly presented below. The purpose is to illustrate various aspects of the entrepreneurial spirit in China, operating in a vibrant market, characterised by brutal competition, demanding customers and strong top-down government control. Common factors in these firms are the speed of their development, and the bold decisions of entrepreneurs who are often well connected to the public establishment.

China's largest food company

A very large firm: COFCO is China's largest food company. It has more than 100,000 employees and sales volume of RMB 200 billion. Founded in 1949, it is one of 53 enterprises reporting directly to the State Council. It has sales volume estimated to be $65 billion in 2015. It has evolved from a trading company of grain and oils to a very large supplier of diversified products and services in the agricultural products and food industry.

Covering the complete value chain, from the farm to the dinner table, COFCO has a range of branded products and service portfolios, including Fortune edible oil, Great Wall wine, Le Conte chocolate, Joy City shopping mall, Yalong Bay resorts, China Tea products, COFCO-Aviva Life Insurance, etc. As an investment holding company, COFCO has five companies listed in Hong Kong, namely, China Foods, China Agri-Industries Holdings, Mengniu Dairy, with which the French food firm Danone signed an agreement in 2013, COFCO Land Holdings and COFCO Packaging Holdings, as well as three companies listed in mainland China: COFCO Tunhe, COFCO Real Estate and BBCA. COFCO has acquired international agricultural products and commodity trading companies – Nidera and Noble Agri in 2014, accelerating its global expansion. In the period 2006–2010, the firm tripled in size.

Food is a concern to people in China after a number of scandals in the supply of alimentation products, most notoriously baby milk. A major reason for these concerns is corruption, which prevents the food control authorities from doing their work properly. This is to be contrasted with Japan, where the food quality controls are strict.

SMEs account for 80% of the jobs in China. The twelfth five-year plan provides particular support to the SMEs in order to develop 'new, distinctive, specialised and sophisticated' firms.

On-line market Yihaodian

As a fast-growing SME, Yihaodian illustrates the very rapid growth of a SME into a major player in the online shopping industry after being founded as recently as 2008. Five years later, the company challenges the leader in this area, T Mall which belongs to Alibaba.

In order to fuel this growth, the firm has rapidly broadened the scope of its offerings. Today, it deals with close to 500,000 orders each day. In late 2014, the firm had 10,000 employees. Everything is monitored via information technology: orders, stocks, managing delivery (there are more than 200 delivery points). In order to finance this growth, the company has welcomed Wal-Mart as a shareholder.

Yue Bao: an example in financial services

Yue Bao is a money market fund using big data from Alibaba, to anticipate the behaviour of customers. It caters to small investors. This fund has met with great success and is expected to rapidly continue growing from its level of $90 billion in 2014. This is unless the government slaps on limitations on its activities, as they now are outside any regulatory framework. This fund could thus be considered to be part of the shadow banking sector of China, which some people estimate to be of the same order of magnitude as the GDP of Germany.

Chinese firms becoming global

There are two main ways for Chinese companies to become global. One is through organic growth and development; this is the path essentially followed by the telecommunications company Huawei, as well as the manufacturer of white goods Haier. Another route is through acquisition.

Geely

One of the first examples of a Chinese company acquiring a foreign firm is Geely. The latter is the tenth-largest car maker in China, founded in 1998. In 2010, it acquired Volvo for €1.3 billion. The two companies are radically different in their history, make up and dynamics: Geely is growing, Volvo is retrenching. Geely thus gains access to technology and a sales network, particularly in the USA.

Geely leaves Volvo functioning as a stand-alone firm. A new Volvo plant has been built in Chengdu and an R&D centre was established in Shanghai. In late 2014 Geely announced that it is investing $11 billion in Volvo to make new models. In the course of 2015 or 2016, cars will be exported to the USA; this will be the first time that a fully Chinese-built automobile will be sold to North America.

Chinese companies are on the threshold of a wave of fairly massive investments in Europe. One of the first purchases of a foreign multinational by a Chinese company is that of the Italian company Pirelli. The acquirer is a subsidiary of the cash-rich state-owned enterprise ChemChina. The proposed price was €7.3 billion. SASAC, mentioned earlier in this chapter, oversaw the process. One may wonder why the manufacturer of 20 million tyres per year is participating in such an acquisition. The main reason may well be just to have a window in Europe and to learn how to do business in the richest region of the world with the most sophisticated, cosmopolitan markets.

In such international acquisitions by Chinese firms, the managers of the buying firm should effectively be warned and guided as to what to do and what not to do, particularly regarding communication, in order to avoid negative reactions. Everything should be done so that ill-perceived practices do not make worse the already mildly xenophobic attitude towards foreign acquisitions. An example of things for Chinese management to avoid is as follows.

Lack of timely communication . . .

A few years ago, a Chinese company purchased a French firm. The negotiations concluded without problems and the deal was officially announced. The Chinese acquirer went back to China, without meeting the representatives of the staff and without any announcement of their intentions about their plans and future activities. This deafening silence greatly alarmed the local French staff and triggered rampant rumours and a huge amount of stress. Everybody feared that there would be massive layoffs. Nothing of the sort happened, but a lot of damage was done to the staff motivation out of negligence and it took a long time to reestablish a good level of motivation and trust.

In their attempts to reach outside the country, Chinese firms sometimes encounter obstacles in their attempt to purchase a foreign company. This was the case when the oil company CNOOC tried to purchase Unocal in 2005. The acquisition was banned by the US authorities and Unocal was later bought by Chevron. Much later, in 2010, CNOOC managed to gain entry into the USA by buying a share in the Texas gas company Chesapeake.

The following example is that of a manufacturer of consumers' products with global reach and excellent value-to-quality ratio. It is active in the highly safety-sensitive area of items for very young children.

Goodbaby

Goodbaby International is an industry leader for strollers, wooden beds and car safety seats for children. Founded in 1989, it is based in Kunshan. It has 20,000 employees in 70 countries. The company is known for innovation with a high level of safety and quality in its products. The company is a partner of several well-known distribution brands in Europe (Carrefour), North America (Wal-Mart) and Asia.

The company has nine global R&D centres, in Bayreuth, Vienna, Prague, Berlin, Dayton, Dongguan, Tokyo, Boston and Kunshan. Goodbaby's innovation allows them to produce over 500 new products every year with suggestions and market research analysis from all of their R&D centres. 'The Chinese department is responsible for the innovation of structure and original design while the design centres abroad pay attention to productization'. The way Zhenghuan Song, President and CEO of Goodbaby Group, has made its R&D department innovative is by making sure that departments in all different countries work as a team and work on separate tasks to ensure that all of their resources are best utilised.

Goodbaby proves that quality and safety can go along with relatively low cost. The firm closely monitors its suppliers; its production control and management systems are relentless, in order to meet the highest safety standards. Its agility has allowed it to respond quickly and efficiently to changing market demands, faster than its competitors, which is what sets it apart as an industry leader. Goodbaby has an extensive, world-class testing centre for its products and is the only company to have received official recognition as a successful laboratory to become an independent testing organisation in China. They have received approvals from SGS, based in Geneva, Switzerland, which is the world's leading inspection, verification testing and certification company, as well as from TUV from Germany. The company has been collecting suggestions and innovative ideas from customer feedback via the Internet.

Goodbaby also has sales staff working directly with maternity and childcare specialty stores, helping it with marketing activities, strategies and in-store promotions. One special characteristic that sets Goodbaby apart from its competitors is its innovative idea to experiment with store designs using a 'model showroom' before implementing them at all showrooms. Goodbaby has a separate showroom for its marketing department where it experiments with the placement of new and older products in order to derive more sales by optimising the space, and to understand what customers are interested in. Once the setting in the model showroom is finalised, it is applied to all stores throughout the country to impact positively on the sales of Goodbaby's products.

'Goodbaby International: A world-class quality model'[2] describes many innovations that the company's management is using to motivate its workers and to achieve high levels of quality and productivity. Goodbaby motivates its workers to be extremely attentive to quality and safety. Part of this is a honour board in the plants.

In conclusion, Goodbaby has set quality standards for its industry. The firm innovates further with its products, processes of management and in its production facilities. China-based Goodbaby has managed to build a strong global brand of quality and customer-orientation, while keeping pace with the constant need for change and novelty.

Becoming global: the art of war?

It is frequently mentioned that in their expansion, Chinese companies are inspired by the 2,500-year-old book *The Art of War*, with thirteen chapters on military strategies and tactics. This is then somewhat analogous to the game of Go (Weihi in Chinese), in which two players seek to secure territory on a *goban* board, with small black and white pebbles. Encircling the enemy would be a metaphor for the way firms go global.

There may be some valid analogy, but let us now only select a few quotes of wisdom from this book:

'In the midst of chaos, there is an opportunity.'

'What enables the enlightened ruler to conquer is foreknowledge.'

'If you know both yourself and your enemy, you can win a hundred battles without jeopardy.'

It is banal, and somewhat sadly simplistic, that the analogy is so often made between military strategy and tactics and corporate management, reflecting the nature of our hyper-competitive world. But it is also often said Chinese management follows this book, as it does not first challenge its main competitor; instead, it attempts to establish a strong position around it, as in the Weihi game. The competitor finds itself encircled before it realises it.

In brief, in China the public sector and the companies are interdependent in a somewhat unique way. In the coming years, this connection is expected to become less dependent. The young generation of entrepreneurs will attempt to operate much more 'hands off' from the government. However, the public sector retains many levers to control firms. Private companies still need the government for a host of things: a licence to operate, loans, buying land and construction permits, as well as a license to import goods.

In Chapters 8 and 9 we will come back to these two key actors, as they engage in dynamics specifically aimed at making China a country for innovation. The patterns followed by these actors in the course of this journey will be documented, in order to provide insights on how China is reinventing innovation by adapting it to the country's idisosyncrasies.

---

[1] One blog on the anti-corruption campaign is as follows: www.anticorruptionblog.com.

[2] This section is derived from a 2015 student paper by Kainth Parame, St Gallen University, Switzerland.

Framework conditions for innovation in China

As in any country, the so-called 'framework conditions' are a set of context elements, such as a country's culture, its overall adherence to the law, interrelated policies, its demographics, etc., which in various and often complex ways favour, hinder, and otherwise influence innovation and inventive conditions in a market. Some framework conditions can be very explicit, such as product safety regulations as articulated by the State Food and Drug Administration that affect drugs developed in the life science industries. In general, these framework conditions include legal aspects, including those concerning Intellectual Property Rights (IPRs), economic policies facilitating entrepreneurship, or tax incentives for financing innovation. These conditions are determined by the actors described in Chapter 4, while the aspects relating to the very important human factor are discussed in Chapter 6.

Demographics, the middle class and urbanisation

One element in the background of the framework conditions is sociological in nature and concerns the fast-growing middle class in China. It is anticipated that by 2022 more than three-quarters of the urban population will be earning between US$9,000 to $34,000 per year. The people in this segment of the population drives the demand for innovative products and services the most, as its dispensable income allows them – for the first time! – to fulfil all those needs and desires that they have only dreamed of for a long time. Many of these people are relatively young, educated and just starting families. Therefore, they have a wide range of spending patterns.

Another crucial element in China's demographics is that of aging. Due to its one-child policy, but also because many choose to focus on economic security before starting a family, too few Chinese are being born to compensate for the many elderly Chinese who now benefit from improved health care and living conditions. The average age continues to increase; soon there will be only few people left of working age to support the elderly or the dependent young. In 2050, China expects to have 480 million people over the age of 60. This has consequences for innovation and product development and redesign. The total amount of goods and services for that population is expected to represent close to one third of the country's economy, opening up what used to be niche markets in China only.

Concerned economists observe that 'China may well be old before it is rich'. They are worried about the so-called 'middle-income trap' that needs to be navigated around by every fast-emerging country. The risk is that China's relative growth will slow down before it has grown enough in absolute terms. This is connected to demographics because large masses of young people usually allow the country to maintain relatively high upward growth rates. To avoid this trap, China is attempting to switch from investment-led growth to consumption-led growth. However, it is very difficult to change the patterns of an economy that has been so spectacularly successful for so long!

Two events had a profound impact on China. First, the Cultural Revolution, which in many ways was an attempt to eradicate anything traditional. This had a terrible impact especially on the basic science and university infrastructure, and it is taking decades to rebuild the tacit knowledge and experience lost. The second influence comes from outside China, through ease of communication and travel. The large Chinese diaspora is also a crucial contribution to the country, especially with regard to innovation and management. We will discuss these in Chapter 6.

Legal aspects

In Chapter 4 we looked at the judiciary apparatus, so let's now look at the broad set of legal conditions. During the period of strictly planned economy, following independence in 1949, all decisions were taken by the political leadership and the bureaucrats, so the legal system had a minor role. This was further diminished during the calamitous Cultural Revolution. Since the 1980s, the legal system has had to rebuild its role – and trust – in Chinese society.

In 1992 the fourteenth National Congress of the CPC made the decision to establish a socialist market economy. It indicated that this must be regulated and guaranteed by a legal infrastructure. The Chinese legislature enacted a set of economic laws, to regulate the actors of the market and maintain market order, while opening the country to the outside world. Many laws have been enacted since, such as Company Law, Partnership Enterprise Law, Law on Commercial Banks, Anti-Unfair Competition Law, Law on the Protection of Consumers' Rights and Interests, Product Quality Law.

Largely as a result of joining the WTO in 2001, China overhauled its entire legal system, including laws concerning companies, antitrust regulations, bankruptcy, etc. It is not our intent to be comprehensive and only a few examples will be highlighted, especially those that affect innovation directly, such as the IPR law, discussed further below.

Intellectual property laws

In its extreme form, corruption fits well in rule-of-man societies, as opposed to the rule of law. The rule-of-man means that the situation-specific interpretation of principles, rules and laws overrides the original, general intention of the promulgation of the law. These interpretations are susceptible to corruption.

China has been trying hard to suppress rule-of-man, creating and rolling out laws and regulations that make it easier for its citizens to do business and invest in innovation. One key piece in this trust towards greater legal stability is China's intellectual property (IP) law. The State Intellectual Property Office (SIPO) was founded in 1980, and the first patent law was promulgated in 1985, then revised again in 1992, 2000 and 2008. The laws are regularly updated. For example, in mid-2015, SIPO invited comments on a new draft of the patent law, which will have to be reviewed by the State Council.

China has joined the Patent Cooperation Treaty (PCT), administered by the World Intellectual Property Organization (WIPO) in Geneva. Upon joining the World Trade Organization (WTO) in 2001, China adopted the Trade-related IP rights (TRIPS), and promulgated an IP legal system consistent with that of the OECD countries. China's laws were thus at world standards. Their implementation and practice, however, were poor and it took a long time for the courts to understand this new field. This reflects the fact that the Western concept of a 'patent' is relatively new to China. However, trademark law was promulgated in 1982 and the first copyright law in 1990.

If there has been an infraction of IP rights, there are two routes available to file an IP suit. The most common approach is an administrative lawsuit via the provincial or city IP office. Alternatively, using a judicial approach, the case begins at the Intermediate People's Court. However, this is fairly complicated and costly, so in close to two-thirds of the patent infringement cases, the administrative route is preferred.

As a rule, courts do not award compensation for lost profits and have been reluctant to award large damages. This may change in the near term, as the *Chint v Schneider Electric* case illustrates. Generally, penalties for IP infringements do not exceed $100,000. Considering how much it costs just to get a patent filed, approved and maintained (easily $25,000 over a patent's lifetime in China) and the money that had to be invested to get to a patentable technology in the first place (a Brody/Berman study puts this number at $1 million for IBM, $7 million for GM, and $50 million for Pfizer), these penalties are no deterrent.

China's patent registration is based on the 'first to file' principle – like almost all countries in the world. Even the US switched to the 'first to file' rule in 2013, although here it is called the 'first inventor to file'. Patents may be filed domestically or under the Patent Cooperation Treaty (PCT). The latter route is open only to invention patents and utility models.

The practice of IP law has been to progressively detach itself from political pressures, especially for litigation between a Chinese and a non-Chinese firm. Considerations of protecting the local company and its jobs were often paramount in the case. This has changed and non-Chinese firms have been winning litigation cases against Chinese companies.

The fairly rapid improvement in effectively implementing IP laws does not primarily come from the pressures of international corporations active in the country, but rather from large, domestic companies such as Huawei, Haier or Lenovo, who want to have the benefit of an IP system protecting their own technology positions. Specialised divisions have been instituted, the first one in 1993 in Beijing.

The Chinese State Council has recognised this issue and outlined a solution in 2008 in its 12-year strategy: 'We should improve the trial system for intellectual property-related cases, optimise the allocation of judicial resources and simplify remedy procedures. We should consider setting up special tribunals to accept civil, administrative or criminal cases involving intellectual property.' In late 2014, special intellectual property courts were established in Beijing, Shanghai and Guangzhou, to deal with IP litigation cases. Thus, innovative firms tend to locate in one of these regions, so that they are in a better position to sue counterfeiters in court. It is estimated that more than 98% of all IP infringement cases in China are now Chinese firms suing other Chinese firms. Once the aggregate inflicted damage on Chinese IP becomes greater than the potential benefit for China (in its various ways!), we expect a significant tightening of the IP practice with benefits for all IP holders, domestic and foreign alike.

As indicated in Chapter 3, Chinese patent law protects three types of intellectual property: invention patents, which are similar to the international notion of patents, as well as much weaker (less inventive) utility models and the even less rigorous design patents. The latter are granted for ten years, whereas invention patents, as in the West, are valid for 20 years.

While invention patents undergo considerable scrutiny during the application process, utility patents are at best checked for domestic prior art and most companies consider it a registration process rather than an application. Despite weaker protection, utility and design patents provide firms with ammunition to sue companies and to extract sometimes large sums of money from them, as is narrated below in *Schneider vs. Chint*. In this sense, these tools are part and parcel of the overall policy of 'indigenous innovation', stressing the importance of China-grown innovations in the economic development of the country at this particular juncture.

Chint vs. Schneider Electric

The Chinese firm Chint and the French firm Schneider Electric are competing in the low-voltage electrical products business. Chint is an industrial equipment company founded in Wenzhou in 1984 (by 2014 it had 30,000 employees and $6 billion in revenues). During the 1990s and early 2000s, Schneider had successfully limited Chint's international expansion with infringement lawsuits (e.g. in Germany and Italy). In a defensive move, Chint sued Schneider in 2006 over a miniature low-voltage circuit breaker used as an air switch in the building industry. This product was protected by a 1999 utility patent held by Chint, which – according to subsequent court proceedings – Schneider had infringed upon with its business in China. In 2009, a Wenzhou-based court ordered Schneider Electric to pay $45 million (RMB 335 million) in damages to Chint. Schneider managed to settle for a payment of $23 million (RMB 157 million) in 2009. Still, the entire lawsuit and the settlement has caused ripples throughout the IP community in China, because it was based on a utility patent, which is perceived to be much weaker than an invention patent, because it was won by a Chinese firm against a foreign multinational, and because of the amount of the claims, damages and the eventual settlement concerned. It is expected to encourage foreign and Chinese firms in China to invest in R&D and pursue protection for the intellectual property.

Counterfeiting and copying

As in most of Asia, copying while improving is a way of life, as suggested by the ancestral practice of calligraphy. It is also a massive industry. China's output of counterfeit goods amount to approximately three-quarters of the world's total. It is estimated that counterfeit goods represent 2% of the total world trade, amounting to $30 billion of imports per year into the European Union and to the USA. Turkey is considered to be the world's second-largest country for counterfeiting. It is estimated that in this country alone $10 billion worth of counterfeited goods are sold each year.

Everything (that sells well!) is copied, products of course, but also brand logos, packaging, service outlets such as fast-food restaurants, complete stores (such as Apple and IKEA stores), hotels, and amusement parks.

Chinese products are not immune to being copied: China's agency in charge of commerce, SAIC, recently claimed that 60% of all products sold online by Alibaba, the world's largest internet sale company, are fake (Alibaba has vehemently denied this assertion). At some point, the Tsingtao beer was copied by Quangmai and sold at 25% below the price of the real beverage.

Building the Shanxi museum with counterfeit anchors

For the world exposition of 2010 in Shanghai, a monumental red building with an inverted pyramid structure was built. To build it, some 50,000 special metallic anchors were necessary.

The builders only purchased 5,000 such anchors from the German firm Fischer. The remainder were bought from a Chinese manufacturer, which closely copied the device. However, the metal used later showed corrosion. This caused concern to Fischer, as guarantee clauses could potentially have resulted in a large claim for damages. The German firm alerted its government. Eventually, the German Ministry of Economics obtained from the Chinese Government proof of the copying, and the guarantee obligations were lifted. The building has since been converted into an art museum.

Copying (which usually refers to patent infringements) and counterfeiting (which usually refers to trademark abuse and impersonation) can result in life-or-death situations. Drugs, for example, that do not contain the proper active ingredients (this is the case in a third of fake drugs) such as antibiotics and cardiovascular drugs are sold on the Internet for a fraction of the official price, luring consumers to believe they are now doing something about their illness while the underlying disease continues unchecked. In some cases, the drugs are actually authentic, but have expired: Unused drugs were bought back from patients or hospitals, stripped of the original package with the accurate expiration date, and then repackaged with a new expiration date. These cases are particularly hard to counter against.

In China, consumer products are particularly ripe for copying, but business to business offerings are copied as well, as are complete systems such as plants, stores, restaurants, and transportation systems. One example is the Chinese high speed train. China bought the rolling stock (and the technology associated with it) from three sources: France's TGV, Germany's ICE and Japan's Shinkansen. Much of the input came from Siemens, as discussed elsewhere in this book. The Chinese partners adapted and fused these inputs into their own brand of trains . . . without incurring any patent lawsuit. Currently, a network of high speed trains longer than that of Western Europe carries 2 million passengers every day. China has announced that it intends to invest more than $5 billion in a high-speed railway linking Moscow to Kazan; plans are underway to connect to several Southeast Asian cities.

Another example is the Siemens-Thyssen magnetically levitated train (maglev), which currently operates – at a loss – between Shanghai Pudong airport and Longyang Station, a station still fairly far from the city centre. Three years after the establishment of the plant to build the train, a prototype maglev was tested by the Chinese Aviation Industry Corporation (CAIC). In the meantime, the Chinese are the only ones who can build maglev trains, as R&D teams in Germany have been disbanded after political and commercial interest in establishing maglev routes in Europe has waned.

In order to avoid copying, large Japanese manufacturers have consistently been reluctant to manufacture their most advanced products outside Japan. In some cases (e.g. Sharp) companies have even moved plants from China back to Japan, in order to be able to protect their most advanced products and manufacturing techniques better.

Another way to protect a product is to have a 'black box' element imported into China. This is a central technical component in the operation of the product and, as it is produced outside China, is more difficult to copy. These black boxes are designed to destroy the inner workings of the core technology in case of attempted reverse engineering.

Reverse engineering of competitor products is a frequent and ordinary practice of companies worldwide. It aims at understanding how the competitive products are made, and, when possible, copying its important elements. In the area of services, copying is common practice, made legal and possible by the absence of IP protection in this area.

In the sector of consumer products, luxury goods in particular, counterfeiting is tolerated, within reason. At an acceptable level, it constitutes a kind of homage to the copied brands and, in some way, acts as an advertisement for the products. It is thus tolerated to a certain extent by the legitimate producers. However, some companies are more adamant in staking their claims, and their goodwill is limited by economic and moral concerns. For instance, the French company Louis Vuitton has had a consistent policy of systematically suing counterfeiters in China.

What is particular to China is that copying is so widespread, so fast and so extensive. It provides jobs, income and profits for local workers (and, by extension, tax revenue for provinces). Much of the product is for domestic demand, but some is also for export. Firms must constantly monitor the market, so as to be aware about the extent of counterfeiting and when quantities of copied goods appear on the market at a level that may threaten their business, brand – or both.

Shanzhai and China's innovation culture

Talking about a country's culture is fraught with difficulty. It is impossible to avoid misleading generalisation and very difficult to extract the most relevant insights. Here we intend to only give a few pointers, which may be useful in the discussion of the topic of this book.

China has one of the world's oldest cultures, at roughly 5,000 years old. 'The nature of men is always the same; it is their habits that separate them' said Confucius. Indeed, Confucianism, Taoism and Buddhism are three key sources of social values in China. The country is extremely diverse and varies from province to province. One manifestation of this is the great range of food specialties and spices available in the regions.

Diversity is good for innovation. A group of clones does not invent much, because there is no cause for disagreement and debate among them. This means that managers must be attentive to have diverse people in their innovation teams, and to change the members of their R&D departments regularly, in order to inject fresh thinking and different perspectives.

However, respect for tradition and authority, 'group thinking', as well as the desire for harmony, are not generally favourable to innovation. This translates into a school system, as mentioned earlier, which does not promote independent thinking, entrepreneurial spirit, or problem solving.

It is also sometimes argued that people in China often have a silo mentality, which prevents them from exchanging information and working effectively across functions – a basic requirement for effective innovation processes. Similarly, a lack of ability for systems integration is noted. This is a serious handicap in conceiving and developing large structures, such as airplanes.

However, the Chinese are extremely good at building and scaling things up, and have superb and rapid execution capabilities. They are very ingenious, which really inspires their engineering abilities, and they are industrious and hard-working. They have a strong sense of loyalty to the group – as represented by their attention to family, or their commitment to a firm – which constitutes some kind of counterpart to the leadership and motivational systems inside Western firms. So, no surprise: it is difficult to paint a black and white picture when it comes to characterising China's cultural influence on innovation!

Sometimes the strongest critics of a system come from within their own country. For instance, Xu Xiaoping is one of China's most prominent business angels and investors in innovative start-up companies. According to him, China still lacks in three important dimensions with respect to a culture of innovation.

First, China focuses on education, at the expense of passion. In his view, many Chinese parents have an obsession with seeing their children pass the Gao Kao, the entrance examination into Chinese universities, a goal that they often elevate over the more natural talents of their children, and the passion and desire that fuel so many innovations in the West. As long as most Chinese seek the safety of a university degree over the innate inclination of the next generation, they are unlikely to come up with the next breakthrough.

Xu also thinks that Chinese companies and their products and services lack 'soul'. He believes the cause of this is China's reliance on engineers whose strengths are in optimising and replicating existing solutions rather than thinking 'outside-the-box'. This would explain the many start ups that clone existing business models and products with just minor variations; highly capable in their own right but lacking the spirit of the original inventor.

Last but not least, innovators are discouraged from exploiting their potential as they are otherwise crushed by large incumbent companies. It is easy for a large firm to just hire another 300 engineers to replicate the product of a start-up company and then scale up their own competing product through their more powerful sales and marketing network, thus effectively forcing the start up out of business. In the West, a large company would more likely acquire this start up and integrate it into its own organisation. This is a very lucrative and honorable exit strategy for many entrepreneurs. In China, however, as long as replication is cheap and easy, start ups will shy away from this kind of Schumpeterian competition.

Perhaps Shanzhai innovation is the new unique feature of Chinese innovation. The practice of 'shanzhai' (literally 'mountain village', evoking bandits in the hills) is intriguing. A Shanzhai product refers to products made outside of government regulations, so relates to the counterfeiting industry. It can also refer to things that are improvised or home-made and generally crude in form and function. Shanzhai products are often cost-innovated redesigns of originals, defeatured of price and cost-driving functionalities and materials. The customer's ability to afford the product is central; clearly no new technologies are involved. However, the combination and configuration of the essential product and its outside components is also innovative. Copying and regrouping is also part of the creation process.

A recent example of a Shanzhai product is the mobile phone provider Tianyu, whose knock off handsets came from nowhere to challenge

Nokia and Samsung within less than two years. Another example is BYD in the car industry, which started as an imitator of competitive products. Initially, its products were batteries for phones, produced at a third of the cost of that of Sony. It then went to auto manufacturing; by 2011 it had 178,000 employees. It is famous for its F3 car model, which is a copy of the Toyota Corolla.

The Shanzhai movement is informal, bottom up, bold and *extremely fast*. It constitutes a pattern that will evolve, but is likely to remain an inspiring element for innovation for future entrepreneurs, as they invest in the realm of hardware items, in a way similar to the makers movement, which we will discuss in Chapter 9.

Corruption

According to the real estate property law, land in urban areas belongs to the collective organisations. Individuals and legal entities, however, may hold a land use right for a period of time and, in certain cases, individuals may transmit this right to their heirs. There are two types of rights: granted and allocated. Only the former can be used by firms as capital contribution.

This right may be terminated, if the conditions of the contract are breached (e.g. the plant does not comply with environmental standards), or if the state expropriates the occupants for reasons of public interest. Local authorities, however, generate large amounts of cash by selling these rights, contributing to substantial real estate bubbles, while creating opportunities for corruption.

According to various rating and research agencies, China is riddled with corruption. Below is a sample list of countries and how corrupt they are. China ranks high in terms of corruption, but it is important to note that this is not a matter of culture, language or Chinese heritage, as otherwise countries such as Hong Kong, Taiwan or Singapore would not be ranked as low as they are. Clearly, other framework conditions also have an impact!

How does corruption affect innovation? Innovation is a risky investment in the future, built on the assurance that economic rents, personal benefits,

Table 5.1 Selected countries and their CPI (Corruption Perception Index) rank and score.

Rank	Country	CPI Score
1	Denmark	92
2	New Zealand	91
3	Finland	89
7	Singapore	84
12	Germany	79
15	Japan	76
17	Hong Kong	74
17	United States	74
35	Taiwan	61
69	Greece	41
69	Italy	41
100	China	36
136	Russia	27
174	Somalia	8

Source: Corruption Perception Index 2014 by Transparency International.

and entrepreneurial profits can be realised for the investor along the value chain. Innovators challenge the incumbent market dominators, an insight Schumpeter had more than 80 years ago. Incumbents are more likely to have the means and the ability to ensure through lobbying and other forms of political deal-making that small companies pursuing innovative solutions do not become too dangerous. As long as corruption undermines transparency and the rule of law, it will be extra costly for innovators to ensure that they can benefit from their investment. This cost can be huge, often completely eradicating any incentives to innovate in the first place. This is why it is so important that President Xi Jinping succeeds with his anti-corruption campaign, although at all levels, not just at the top as exemplified by the 2015 indictment of Ling Jihua, aide to former president Hu Jintao. However, it is impressive how much innovation China has already attracted and generated, despite its poor corruption record. How much more would be possible if China had this problem under control!

Ease of doing business in China

Concerning the 'ease of doing business' and following Western criteria, according to the World Bank (see: 'Doing Business 2015'), China comes out ninetieth. Singapore tops the list and China is somewhere between Kuwait, Namibia and Paraguay. The report highlights two improvements in this area: (1) China made it easier to start a business by doing away with the requirement of minimum capital; and (2) China made paying taxes easier by developing electronic filing and paying taxes.

Starting a new company in China remains a very fastidious affair, which requires jumping through various bureaucratic hoops. No simplification is in sight. This is to be contrasted with electronic registration of new businesses available in Chile, Great Britain and Singapore. Such online services are rapid and inexpensive for the entrepreneurs, while avoiding bribery. In Singapore, for example, the fully computerised registration process means that a company can be incorporated within one or two days, under normal circumstances. Hong Kong, a Special Administrative Region under Chinese authority, has similarly efficient registration processes. Compared to other countries, especially those doing well in the 'ease of doing business' ranking, China does particularly badly in terms of 'dealing with construction permits', which is somewhat counterintuitive given the construction frenzy in China, protecting minority investors, and starting a new business. China does comparatively well with respect to 'enforcing contracts,' which again is somewhat counterintuitive given the complaints about the legal context in China, and in 'registering property'. However, even in those two dimensions China is barely within the top forty. One aspect that was improved in this context was the high share that intellectual property was allowed to take as part of incoming registered capital in start ups. Especially when entrepreneurs have little capital, the technology they invented should account for something, and this has been rectified in China in the last years.

Financing innovation

While China runs a high trade surplus and has rich foreign exchange coffers, much of this capital is not easily and readily available for investment in innovation, especially not for SMEs and start ups who only have a small share in the benefits of international trade.

In China, SMEs are defined as having 300 to 1,000 employees and RMB 20 to RMB 400 million in revenue for medium-sized companies, and fewer than 300 employees and less than RMB 20 million for small-sized companies. 99% of all companies are SMEs, providing 75% of all jobs, and contributing 58% to China's GDP.

SMEs need any help they can get. In China, they are in a difficult spot competing against large private enterprises that are multinational or have strong links to the government (SOEs), or they are up against small entrepreneurial start-up companies that have more flexibility. Ninety per cent of all SMEs stated that they were facing financing difficulties. Banks are often unwilling to extend credit lines to SMEs as they have less collateral to mortgage against, and they operate under greater strategic risk. Most banks in China are state-owned whereas SMEs are not; banks prefer lending to other SOEs with whom they have more in common, including the all-important information network operating behind the scenes. SMEs are more work for investment managers; they are not backed up by the government in case they fail.

Banks are still the greatest source of funding for SMEs though, as alternative sources of capital are even worse. The stock market is unreliable and has been fairly unproductive for many years in China. Venture capital (VC) prefers to invest in high-growth start ups, namely in underserved markets or sectors that can become disrupted or disintermediated. Thus the government has set up several programmes and funds aimed at SME development and innovation over the years.

The SME Innovation Fund was established in 1999, with a budget between RMB 500 million to RMB 1 billion. The SME Development Fund, established in 2011, has a budget of RMB 15 billion. Both funds provide financial support to high-tech start-up firms or SMEs with the plan to accelerate technological innovations. In addition, back in the 1980s, the government has created technology commercialisation initiatives such as the TORCH programme, which encourages investment in high-tech investment zones and about RMB 1.3 billion of investment in more than 2,000 projects during the eleventh five-year-plan alone, or the SPARK programme, which focuses on the development of rural economy through science and technology.

Venture capital in China

Venture Capital (VC) funds were created in the mid-1990s after it had become clear that government programmes, such as TORCH, would not suffice to fund all promising start-up opportunities.[1] Government VCs (GVC) were set up first, then university VCs (UVC) and, in the late 1990s, corporate (CVC) and foreign VCs (FVC) came to China. Over the last ten years, VC investment in China has averaged about $5 billion, which makes China the third-largest VC market in the world after the US with about $35 billion and Europe with about $7 billion of investment annually on average. In 2014, China's share of VC as percentage of GDP was 0.12%. The EU average is 0.277%, more than twice as high. However, China is similar to Italy (0.138%) or Romania (0.137%). GVCs have been replaced by so-called 'guidance funds' (i.e. government investments into private VC funds with explicit missions ('guided investments') for certain types of investments).

In 2015 the China State Council announced a $6.5 billion venture capital guidance fund, to invest in early-stage and pre-revenue companies and technology incubators across the nation. These companies, often managed by engineers and scientists, require a fair amount of support and nurturing by their investors. These funds are privately managed and the proceeds are split between the government (not just the central government but also at county, municipal and city level) and private management entities.

For a long time China's VC industry lagged behind the potential everyone saw in it. Many VC funds did not follow proper due diligence principles or invested capital outside the start-up scene; there were inappropriate reporting standards and general lack of transparency. Many VCs were first-time fund managers and lacked experience and business acumen, while government VC funds were often managed by government bureaucrats. Many start ups were led by first-time entrepreneurs and lacked experience and advisability. Chinese VCs tended to focus on later-round deals with single large investments that promised bigger one-time cash-outs rather than the riskier earlier-stage investments that require greater understanding for underlying fundamentals. Making VC work in China was more difficult than in the more transparent and efficient markets such as Europe, Israel and the US. However, in the past three to four years, co-investments of local VCs with foreign VCs operating in China introduced global best practices, which resulted in very large deals of Chinese enterprises, providing significant returns to their investors.

These positive factors are now visible: The latest numbers for 2014 indicate that China has overtaken Europe as a VC market for the first time on the back of some major large-sized deals, pushing the size of the market closer to $15 billion. The global VC industry has reached $86 billion, second only to its peak of $116 billion in 2000.[2]

A new class of domestic CVCs have emerged, led by firms such as Alibaba, Tencent, Baidu and Xiaomi. Not only do they pick up any remaining slack left in co-investing (with domestic and foreign VCs) in local Chinese start ups, they also have started to make significant investments in overseas start ups, for instance in the US. The purpose for these foreign investments is twofold: (1) access to technology that can be brought back to China and is embedded in the corporate solution offering; and (2) significant amounts of equity investing into more later-stage companies, in order to gain market access overseas. 2014 saw the first time that a US start up, Snapchat, received investment from two different Chinese corporates (CVC), albeit in different rounds.

All these are initial and positive signs of a gradually maturing VC and CVC industry. The trends are exciting and if there is any lesson to be learned from the success of Silicon Valley, it is that VC and private angel investments make a huge difference in innovation and job creation and are accelerators to increase the national competitiveness by moving from 'made in China' to 'designed and created in China'.

The extremely complex process of moving technical know how to a place in an organisation where it results in new activities and jobs includes technology transfer. The latter involves several channels and is the object of the following section.

Table 5.2 China Now Number 2 in Global Venture Capital.

Country	VC Investment (US$) 2014	Number of Deals in 2014
United States	$52.1bn	3,682
China	$15.5bn	740
Europe	$10.6bn	1,460
India	$5.2bn	260
Israel	$1.9bn	196
Canada	$1.4bn	169
TOTAL	**$86.7bn**	**6,507**

Source: Ernst & Young, 2014

Technology transfer capabilities

Technology transfer describes the process by which research results are commercialised by creating new or more competitive activities, and jobs. The research work is usually carried out in a university or a public research organisation, and increasingly also in the R&D laboratories of private firms. The transfer often concerns technology and knowledge; we will use the term 'technology transfer' for short.

There are three main channels for transferring technology: collaborative research, licensing and spinning out a start-up company. In addition, a powerful transfer is made by students graduating from university and bringing their knowledge to the company which hires them.

Collaborative research involves a laboratory carrying out a research project on behalf of a private company. This is sometimes called contract research, and organisations such as Battelle in the USA or Fraunhofer in Germany are famous for it.

Licensing involves granting patent utilisation rights to a firm. This requires a reasonably strong patent position. The transfer may be done to the benefit of a start-up company, or to a more established firm. Usually licensing fees or royalties have to be paid, often in the form of a percentage of revenues or profits or both.

Commercialising research results by founding a company, based on these results, is the most complicated, as it requires solid business and managerial skills.* Established companies have the administrative and legal know how to do so fairly easily, but they usually lack the entrepreneurial attitude of those most familiar with the technology. University spin-offs usually have the opposite problem.

China is intent on developing technology transfer activities, because they represent powerful ways to draw on public research to create new activities and jobs. Several successful examples whet the appetite for more, as for instance the origin of Legend (later called Lenovo), which came out of an institute of the Chinese Academy of Sciences.

Technology Transfer Offices (TTO) have been created in numerous universities for this purpose. The main limitation for them is the availability of experienced and knowledgeable personnel. Staff in the TTO have a complex job; they must deal with technical, business, contractual and people issues in an effective way. Such staff are in short supply even in the USA, Europe and Japan, where a string of TTOs were created following the successful examples at Stanford and the MIT. Management training and traineeships constitute ways to mitigate this scarcity. Also, business schools should be moving to take more on this issue and helping move technical results to the market much more than they do currently.

All countries in the world are making efforts to improve the effectiveness of their technology transfer activities. China has some reasonably well performing TTOs, but they are rare and their development takes considerable time. Peking University and Tsinghua have such offices, paired with incubators, for helping to grow start ups emanating from the universities. Of course, these universities can tap into very rich entrepreneurial and relatively advanced technology districts, many of which were recognised as special economic or high-tech development zones.

It is estimated that 200 million Chinese people speak English. In proportion of the total population, this is a higher percentage than that of people speaking Mandarin Chinese in Western countries. In

* More information on technology transfer is given in books in the Bibliography, in particular *From Science to Business*, by one of the authors of the present book.

fact, there is a strong imbalance in the flow of knowledge between the Chinese and the non-Chinese world. The amount of text translated into Mandarin is 10 times more than that translated from Chinese. A similar situation is encountered in the case of Japan.

Economic and technical development zones

Starting with the first economic and technical development zones (ETDZ) in 1984, in Shenzhen, near Hong Kong, in the province of Guangdong, the number has increased to 219 today throughout China, along with 129 high-tech development zones (HTDZ). They are primarily designed to attract foreign direct investment (FDI), with tax incentives for investing companies. Such investments include technical development centres and R&D units.

The mission of these zones has always been to facilitate China's opening up and make it easier for foreign companies to set up manufacturing and R&D operations. After an initial batch of economic ETDZs set up along coastal provinces, a second batch was approved mostly at the state level in the period from 1990 to 1993. Starting in 1994, the amount of FDI into China began to be substantial. Changes in China's economic zones and industrial areas have been characterised by gradual steps and experimentation. They evoke the pragmatism of the top leadership, while the role of the state has been proactive and acted as a facilitator.

Such zones are located on the outskirts of large urban areas and are used by firms. One of the better known zones is in Suzhou, west of Shanghai. Driving along that zone, one sees numerous plants of multinational companies headquartered in Europe and North America. A large sign proclaims a phrase from Deng Xiao Ping: 'Development is one immutable truth.'

The China-Singapore Suzhou Industrial Park was created in 1994 as a joint venture between China and Singapore. In the 1980s, Singapore served as a model of successful economic development for Asia and the Chinese government wanted to benefit from the managerial skills of Singapore. In the late 1990s, however, the substantial losses suffered by this project created a scandal in Singapore. The park covers 288 km² and includes an area for higher education, Dushu Lake Town, which comprises 25 universities and colleges. In 2013, there were 80,000 students enrolled in these universities alone.

Future special zones are no longer likely to focus on attracting FDI, but rather essentially to offer areas for further experimentation. One example of this new zone is the Special Zone in Shanghai, announced in 2013 and mentioned in Chapter 8. One purpose of that zone is to test reforms in the financial sector.

However, the most reputed zones today are still in Beijing (e.g. Zhongguancun, also known as China's Silicon Valley), and Zhangjiang High-Tech Park in Shanghai-Pudong, one of the hubs for foreign R&D in China.

Infrastructure

In 2011 China opened the world's longest bridge in Shandong province, near the city of Quingdao. The Jiazhou Bridge is a structure with 26 km over water. As the visitor travels across the country, the feeling is inescapable that the Chinese seem to have some pride at building structures and edifices.

The development of the country's infrastructure is a high priority for China. It already has a good infrastructure on the coastal regions, when it comes to energy production, rail, road and airports, as well as telecommunication. More investment now goes to the central regions of China (as part of the 'Go West' policy which also affects many foreign investors in China). Compared to India, China's infrastructure has a clear competitive advantage.

China has steadily invested in infrastructure, by some estimates twice the share of GDP compared to countries such as the US or Europe. It also uses such investments to stimulate the economy. In 2014, China announced a $100 billion investment in infrastructure projects, mainly for roads, ports and airports (it plans to have 240 airports in 2020). Early in 2015, roughly 300 infrastructure projects were accelerated with the same objective.

Such building sprees account, in large part, for the large debt (roughly $3 trillion) of the provinces. Needless to say, as in the West, some of these infrastructure investments are questionable, such as airports with one plane a day, or a railroad going nowhere. On the other hand, with all the construction activity being carried out at a frenzied pace, several companies have developed innovative ways to build very tall very fast.

Mini Sky City, a 180,000 square metre building with 57 storeys, was completed in just 19 days using a new modular pre-fabricated construction method. The plan had been to build 97 storeys, but local air-zoning regulations required the building to stop short. Broad Sustainable Building, the company behind this technology, has plans to go one better: It promises to build the world's tallest building, Sky City in Changsha, an 848 metre skyscraper with 200 floors and more than 1 million m^2 of usable space, in just 210 days – 120 days for the pre-fabrication of the construction elements, and 90 days for assembly. It will be built with some of the best energy efficient technology, and will be able to resist even the strongest earthquakes. Even though the company has already erected 20 buildings in China using this technique, there are still some doubts that a building of that height would be able to withstand wind loads. At a cost of $1,500 per square metre, it is just one-third of the construction price of the Burj Khalifa (most construction material and work are required to be sourced from Changsha to benefit the local economy). As always, engineering will need to prove what it is capable of; innovation has always been about testing the boundaries of the possible and desirable.

It is expected that in coming years, with annual growth rate in the 7–8% range in 2014, as compared to double-digit growth for most of the early 2000s, China will continue to support infrastructure investments to boost the economy. This includes an ambitious urbanisation plan to build new cities.

Such plans also concern international projects such as the $40 billion Silk Road Fund for land and sea infrastructures. Another is the Asia Infrastructure and Investment Bank (AIIB) which was officially launched in June 2015 with 57 member countries, including close allies of the USA, such as Great Britain, France, Australia, despite the early and clear opposition of Washington to the project. The USA,

Canada and Japan have declined. Countries join as they can expect preferential treatment in trade and contracts for infrastructure projects worldwide. AIIB has a capital of $100 billion, of which 30% are owned by China.

Specific infrastructure projects may very well have a geopolitical motive. In this respect, we may anticipate the follow up to the ongoing Spratly project. The Spratly islands are on a crucial sea route in the South China Sea. It is the object of construction work aimed at filling in some of the sea in order to enlarge the islands and build an airport and to claim supposed oil and gas reserves offshore. These islands are, however also claimed by the Philippines, Brunei, Malaysia and Taiwan, another reminder of the many points of tension in Asia.

The telecom infrastructure

In 2015, it is estimated that well over 900 million handsets are operating in China; these include smartphones and tablets. In 2014, a new joint venture was formed, the China Communications Facilities Services Corp. Its purpose is to build, maintain and operate wireless towers and infrastructure. It is the result of the coming together of three state-run operators: China Mobile, China Unicom and China Telecom. The resulting joint venture has a $1.6 billion capitalisation to provide added horsepower to the continuing development of the telecommunications infrastructure.

The latter already includes three million kilometres of fibre optic cable (as of mid-2015). The number of landlines is decreasing fast, while it is estimated that more than 150 million devices are connected in a 3G system. In this area, China's own TD SCDMA standard, a technology in which considerable effort, capital and pride was invested during the early 2000s, has not been a success.

Steal or innovate?

Why should China innovate, some people say, when it can just steal from other countries? We hope that this chapter has cast some light on the improving framework conditions in China. China cannot steal technology anymore, at least not to the extent it used to (and many other countries perhaps still do), because it is trying to turn the corner from being a developing country to becoming an industrial country, where the legal framework and context favours investment in innovation and protection of IP, and because it is reaching a level of technological sophistication where stealing itself is increasingly expensive, yielding only little competitive advantage any more. Probably there will be pockets or even regions in China where copying and patent infringements will survive for some time to come, but overall, the only way forward is to honour IP rights both for foreign and Chinese companies. This 'way forward' implies significant changes in how China is managed overall, especially in terms of governance and enacted policy and administrative control, as otherwise it will limit China's potential for innovation significantly.

One of the framework aspects in which China has made remarkable progress is its intellectual capital. This concerns the human factor, as competence, talent and motivation constitute absolutely crucial ingredients for the success of the innovation process. It is the subject of the next chapter.

---

[1] We are indebted to Dr Martin Haemmig, a leading expert on global venture capital and China entrepreneurship, for his valuable contribution to this disscussion.

[2] Ernst & Young (2014): 'Adapting and Evolving – Global venture capital insights and trends 2014'; Ernst & Young (2015): '2014 Venture Capital Review'. Available at: www.ey.com.

Chapter 6

The human factor: a crucial element in innovation management

By far the most important factor of success in any human endeavour is the talent and motivation of the people involved. This is true of any activity in firms, non-profit organisations, and governments alike. A poor organisation with excellent people is likely to perform satisfactorily; a superb organisation with poor and demotivated personnel will fail. Talent and motivation are even more crucial when it comes to the intricate and complex issue of managing innovation, that is: the highly complex process of turning an idea into commercial success. 'Experts', consultants and business school educators promote their own particular theories and 'frameworks' – usually old wine in new bottles – to explain how to improve the innovation process in firms. They do so, instinctively, in an attempt to justify their existence. There is no panacea, no magic shortcut for effectively climbing the arduous path from idea to commercial success.

Talent and motivation

If talent and motivation are the key ingredients in managing innovation effectively, then management has the opportunity to create substantial value for their organisation. First, managers do this by hiring the best people for the job. This is crucial for success.

Second, they must take the time to properly coach and develop their staff to acquire more business sense and managerial practice. This is particularly true of engineers, who must add process skills to the contents of their technical knowledge. Technology-intensive innovation being a multifunctional, multi-actor process, the aim is to have R&D personnel effectively interact with all personnel involved – especially outside R&D itself, i.e. patenting, design, manufacturing, marketing, etc. – as well as customers if appropriate.

Last, and by no means least, management must try to motivate their staff. This involves a mix of excitement and trust, a very important and too rare ingredient in firms. All this means that managers need to be close to the action, while remaining very demanding. The most appropriate style of management to foster these characteristics is briefly described in the following section.

Managing by walking around

Technical knowledge workers need the supporting style of 'management by walking around'. Such a style is best suited to maintain an ongoing dialogue, while developing staff to acquire more of a business sense and an entrepreneurial perspective. This should be done within a coaching perspective.

Because of the importance of the scientific content of their work, R&D professionals need frequent exchanges (not only with their peers), in order to exchange ideas about technical matters and the latest publications or conferences in their field. They also need frequent contact and support from their line managers due to the inherent uncertainty of their work. An open door policy and a 'walking around' style of management are best suited to ensure this.

Managers regularly and informally seek information on the advancement of projects, in order to be able to provide guidance, contribute timely technical input as well as appropriate business intelligence. These contacts take place in the laboratory, in the offices, or in neutral spaces such as hallways, the library, the cafeteria or by the photocopier, occasions for serendipitous meetings. Typically, a manager of a unit including 35 staff carrying out roughly 50 different projects thus spends one hour every day keeping her/his 'ear to the ground' in this way.

The manager must sincerely show empathy towards the 'masters of the craft' researchers, while understanding the substance of the project. For this reason, it is rare that a non-technically trained person is supported very long by the staff of the unit. It is advisable for a manager to leave an R&D unit, go to a business job and come back later to the technical unit. Such a job rotation is powerful in allowing those working in a technical unit to acquire some business sense.

Despite the availability of millions of engineers, executives from non-Chinese, global companies have consistently indicated that their top issue in China is to find and retain appropriate Chinese talent. In this respect, non-Chinese corporations are slightly at a disadvantage: their rules prevent them from offering their employees the generous 'perks' (e.g. a car, an apartment), that Chinese firms can provide.

Managerial constraints within the firm constitute an important part of the human factor in making innovation successful. The considerable contribution in this area from multinational companies operating in China is discussed in Chapter 7. Many elements outside the firm are relevant to the human factor in the innovation process. A first element is education. Although this is not the place to review this area in detail, it is highly relevant to look at education as a provider of a solid grounding in knowledge, as well as a perspective for entrepreneurship. A basic building block is constituted by the school system.

Education as a promoter of innovative and entrepreneurial spirit

By and large, China's high school system is deemed to excessively rely on memory and strict disciplined learning, with little encouragement of individual initiative. A metaphor for this are scenes of hundreds of children, carefully lined up in the school yards to do gymnastics. Conforming to social pressure, extreme obedience to authority and rigid discipline are elements opposite to what constitutes an innovative and entrepreneurial spirit:

- Large student bodies, with mostly young inexperienced teachers, lead to written exam testing rather than personal assessments.
- Lack of team work at university level.
- Confucian value of respecting teachers leads to non-challenging the status quo/established view, loss of learning.

- Focus on graduating fast (getting into the university is the hard part, graduating is easier) leads to undifferentiated skillbase (no work experience, no travel . . . generally less capable graduates).

However, Chinese students know how to cram in vast amounts of material in a short time, which is somewhat useful for ramping up their skills in later training and in professional activities.

Mandatory schooling

In 1949, China had one of the world's highest illiteracy rates. Close to 80% of people had trouble reading and writing. The PRC government made education a top priority and rapidly built schools. Today, literacy rates are in the neighbourhood of 90%.

Young Chinese must attend school for nine years, according to the compulsory education law of 1986. Primary school stresses Chinese and maths. These 'big two' disciplines represent 60% of the school hours. Admission to senior high school is the subject of an examination, the Zhangkao.

Only in the best schools is the examination very competitive. Such schools are achieving good quality learning. The PISA study compares the abilities of high school pupils in a number of countries.[1] This Programme for International Student Assessment (PISA) makes a remarkable exception for China. The study has a bias, as it only considers selected schools in Shanghai, calling the sample 'Shanghai-China'. It is a bit like saying that for that study, France had decided to only take into account a small sample of its best high schools. The media, when talking about PISA, usually do not mention this 'Chinese exception', only celebrating the outstanding performance of Chinese students.

Students from 'China-Shanghai' indeed secure top ratings in mathematics, followed by Singapore, Hong Kong, Taiwan and Korea. Performance in reading and science is world class. The question is: what about China's high school education, in general? Indications are that its level is low; discipline and learning from memory are favoured to the detriment of team-building, initiative and entrepreneurial spirit.

Expanding and improving the quality and job potential of vocational schools has been a long-standing objective of the ministry. In China, this type of education has a high dropout rate and is not perceived to lead to employment; it is not highly regarded by the general population.

In 2014, roughly 15 million pupils graduated from the secondary educational system. Building on the foundation of high school education, China's universities cover a wide spectrum. To enter the university system, pupils must take the Gaokao, or National University Entrance Examination. The grades obtained by the pupil will decide to which university he or she may go.

Higher education

In 2015, it is estimated that more than 11 million students are enrolled in some 1,500 colleges and universities in China. In the Beijing area alone there are 83 institutions of higher education. Of these, 34 are affiliated with the central government, while 49 report to the municipal government.

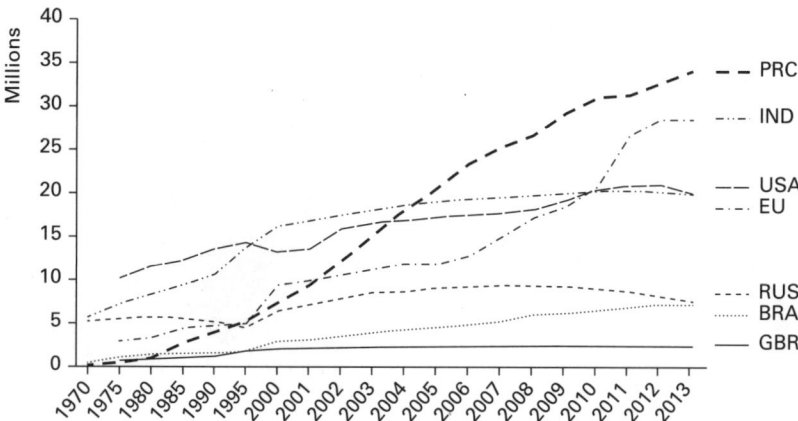

Figure 6.1 Total tertiary education students (in millions) in selected countries.
Source: WorldBankData (interpolated where data missing).

Close to 7 million students obtain a degree every year from China's university system. Thirty one per cent of undergraduate degrees are in engineering, as compared with 5% in the USA. By 2030, China anticipates having roughly 200 million college graduates.

In the area of science and technology, China has 11 million undergraduate students and 700,000 graduate students. The latter number is considerably higher than those of the European Union and the USA. In 2013, 1.5 million students graduated with a degree in science and engineering.

When it comes to research and doctoral work, China has some way to go before it develops a vibrant 'academic' environment. Doctoral candidates often choose their PhD supervisors more for their status than for the topic of their dissertation. The student to supervisor ratio is very high at 5.7:1. China graduates about 64,000 PhDs per year (the US about 50,000). On average, PhDs have a lower employment rate in China than Bachelor graduates, signalling that China's science output may be a bit ahead of its industrial needs.

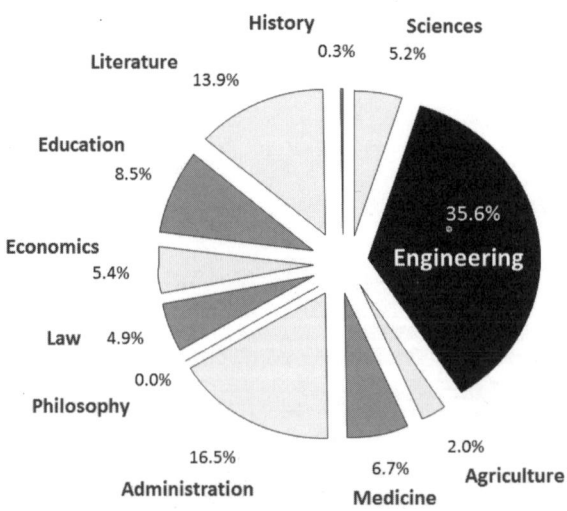

Figure 6.2 The disciplines selected by Chinese students.

In all, China's output of engineers and scientists is impressive in numbers and of fair overall quality. It is definitely a solid basis on which to develop capabilities for developing technology-intensive innovations.

China's diaspora as a source of talent

For centuries, Chinese merchants have been present in the South China sea and the Indian ocean. In the nineteenth century, a shortage of labourers in the USA, Canada, New Zealand, Brazil and Western Europe attracted large numbers of Chinese immigrants.

Between 1978, roughly when China's economic development took off, and 2000, the number of overseas student returnees has gone up from 600 to more than 60,000 every year. The nickname for the returnees is HaiGui ('sea turtles'), i.e. marine turtles that come back to lay their eggs on the beach where they were born.

Between 1978 and 2013, it is estimated that close to 1.5 million Chinese have returned from abroad.[2] The China Youth Returnee Association is one of the more powerful non-profit organisations in China.[3]

The Beijing region attracts almost half of the returnees. Close to a quarter of them choose to start their own company, taking advantage of the incentives mentioned above. In 2013, an estimated 413,000 Chinese went to study abroad, while 353,000 returned from overseas studies.[4]

This diaspora of experienced Chinese, who have spent many years studying and working abroad, is a great asset for the country. In the past, Taiwan has made good use of a similar situation, attracting overseas Taiwanese, who worked on the US West Coast by building industrial and science parks similar to those in California. As will be seen in Chapter 8, much was done by the Chinese government, supported by the provinces, to attract and retain 'returnees'. Their homecoming has been crucial to the development of certain industries, pharmaceuticals and life-sciences in particular. Totally different from traditional medicine, this Western industry has been forged, not only because it is based on Western science, but also as a result of the rules for drug development, clinical studies and the role of drug agencies, such as the FDA-Food and Drug Administration and EMA-European Medicine Agency.

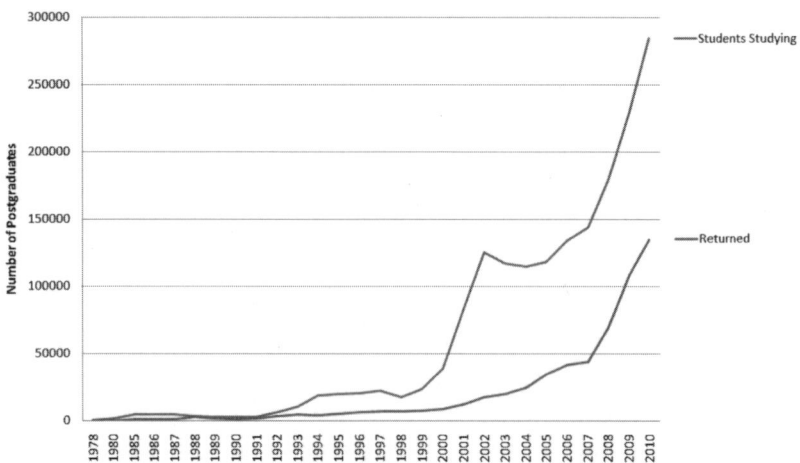

Figure 6.3 Annual flow of returnees going back to China.

Source: China Statistical Yearbook, multiple years.

Looking at the flow of trained people into China, it is anticipated that by 2020 it will have the highest fraction of students studying abroad. Numbers are estimated at 500,000, representing 10% of the world's population studying outside their home country.

A returnee role model

Zhang Xin is a business success story. In 1995 she co-founded SOHO China, now the largest commercial real estate developer in Beijing. In 2014, Forbes ranked her as the 62nd most powerful woman in the world. She is a returnee who before had studied at Cambridge University and worked at Goldman Sachs in London.

She returned to China in the early 1990s when China started opening its doors to foreign investments. She realised that commercial and office buildings in the Shanghai and Beijing regions was going to be good business. SOHO China, incorporated in the Cayman Islands, usually calls on internationally renowned architects to design and build its commercial buildings.

Another initiative to attract talent from outside back into China is the 'thousand talents campaign', launched in 2008. It aims to attract skilled professionals and university faculty, particularly in scientific areas where the needs are acute. One such area is aircraft design and manufacturing for the efforts of the firm COMAC to build the civilian jet engine C 919. This airplane is similar to the Airbus A320-321 and is due to be launched commercially in 2017–2018. The COMAC company is discussed further in Chapter 9.

As of early 2014, more than 3,000 highly qualified individuals have come back to China through this plan. The selected persons must be younger than 55 and either be entrepreneurs or have highly specialised skills. They receive a one-time subsidy and must commit for a minimum of three years, but could be visiting part time from an international base.

Managing the executives

In the important task of selecting the appropriate top leadership in firms, the role of the central government as 'control tower' for the entire country is vividly illustrated by the Communist Party's central organisation department ('Zhongzubu'). This department nominates the roughly 4,000 leaders in the country. These include political personnel and public service, as well as the top executives of the SOEs (State-owned enterprises), presidents of universities and key persons in the media. Housed in a discreet building near Tiananmen Square in Beijing, the organisation is copied from the former USSR's 'nomenklatura' ('list of names'). Senior Communist Party officials occasionally use their status to influence key appointments, but, in attempt to avoid cronyism, the organisation regularly rotates people to various posts. For example, Chen Deming was named deputy governor of the province of Shaanxi after being mayor of Suzhou and president of the administrative committee of the large industrial park in Suzhou.

The Communist Party constitutes a powerful tool for implementing policies, ensuring some degree of consistency across the country. Overall, it is difficult for observers, Chinese or not, to reliably 'read' the real policy trends. 'Realpolitik' sometimes trumps the objective of reinforcing China's innovativeness by proper use of an improving scientific and technical base. In fostering innovation, however, there are strong incentives in place; they concern the nature of the character of the government and the laws and regulations, as discussed below.

Continuing education

By the year 2020, it is estimated that China will have 350 million working people who will have taken some kind of educational course during their employment.

There is already a large number of such courses. In particular, courses are offered to entrepreneurs on the sites of science parks, and are often offered to 'returnees' at subsidised rates. This indicates that massive continuing education is taking place in China, including in management development. Indeed, carefully crafted courses can be very effective in developing young entrepreneurs with the appropriate mindsets and competences for management in general, and for managing innovation in particular.

The leader in this area is the China Europe International Business School (CEIBS). This institution is considered one of Asia's best business schools, founded by the European Commission in 1994 as a joint venture with the Shanghai municipality. CEIBS currently has an international faculty of close to 70 members. Its 20-month flagship Executive MBA programme 'Global EMBA' has more than 1,000 participants, making it one of the largest in the world.

As China's firms become increasingly active outside the country, the need for knowledge of international business practices is rapidly growing. This need can be partially fulfilled by well-targeted executive education programmes. The Chinese market must be actively developed, however, since Chinese firms seem somewhat reluctant to pay the price for immaterial services, such as consulting, advertising and executive education.

Developing the capacity for entrepreneurship and for managing innovation must remain a high priority. This can be done at the various stages of the educational curriculum, as is practiced in many countries. The power of successful and bold role models for entrepreneurs is paramount. Their testimonials in the classroom are a powerful tool for teaching and inspiring students. However, the entrepreneurial teams of young companies must also be developed. Effective 'science parks' are probably the best place for putting programmes in place that allow experienced entrepreneurs to act as coaches of start ups, as they develop their activity, thus transferring their experience and business knowledge to the entrepreneurial teams.

China's demographics represent another challenge. Between 2015 and 2050, the UN anticipates that China's ratio of working-class population (aged 15 to 64) to dependent population (aged below 14 and over 65) will precipitously decline from 3 to close to 1.5 (i.e. at similar levels to those found in so-called 'developed economies').

Thus, it will be even more crucial to have a good fit between people's skills and the needs of the labour market. The objective is to achieve the balance and low unemployment, enjoyed by small Switzerland, in China's large and rapidly evolving economy.

Mindset and 'management culture'

Starting in the sixteenth century, the Jesuits set out to build a channel for contacts and understanding between China and Europe. Matteo Ricci, who arrived in Macau in 1582, was one of the first non-Chinese to live in Beijing. He translated Confucian classics and wrote about the country and its people.[7] Before him, Francis Xavier had this to say about Chinese people: 'China is an extremely big country where people are very intelligent and who has many scholars ... The Chinese are so dedicated to knowledge that the most educated is the most noble in their writings.'

Entrepreneurial spirit is the companion of innovation. It provides the energy to power the innovation engine. Thus, maintaining the momentum from 'made in China' to 'created in China' requires that a sufficiently large fraction of the population has with the entrepreneurial passion to turn market-oriented innovations into commercial success.

It is not clear what percentage of the population this fraction should represent, and the quality of entrepreneurs is more important than their number. It is for all to see the power of role models, as personified by successful entrepreneurs such as the notorious Jack Ma CEO of Alibaba. It is great to see the impact such a highly successful icon can have; even more so in a Confucian country, which gives a prominent position to strong leaders.

In China, the younger generation of urban people may not be so permeated with traditional 'Confucian values'. They have, however, considerable entrepreneurial energy. This is encouraged by an environment and government leadership relentlessly supportive of innovation as a key engine for creating new activities and jobs.

Technology-intensive innovation requires high-quality scientific and technical personnel. This is necessary but not sufficient, since the complete innovation process from idea to market calls on contributions from almost every part of a firm. As mentioned earlier, the innovation process is energised by entrepreneurial spirit. It also must be inhabited by the proper mindset and way of doing things. This is broadly described as 'culture'. Let us look at some of the aspects of the typical ways of managing.

Mianzi and guanxi

This section does not intend to provide a detailed and exhaustive discussion of the specific aspects of managing 'the China way'. A small number of pointers will be given below, with the obvious risk of generalisation.

Two important interconnected notions are frequently encountered in Chinese management: *mianzi* and *guanxi*. It is difficult to find a good equivalent to *mianzi*, which conveys the notion of reputation and social status, as well as saving face. Generally, Chinese people do their utmost to avoid looking bad in public. In the West, this is a mostly individual logic; in China, saving face is also aimed at maintaining harmonious relationships among a collective group of people.

Related to it is *guanxi*, which means relationship. It indicates the extent and quality of a person's interpersonal ties. Since maintaining face is important, a person will not take advantage of somebody in her/

his *guanxi* network, thus establishing a circle of trust. Central to the Chinese mind is the idea that 'we are born connected', which is probably a reason why the Internet is such an obsession for young urbanites. At the firm level, it is crucial for an entrepreneur to build a network; most Chinese leaders and entrepreneurs are outstanding at creating such networks. It is often said that, in China more than in other countries, people do business with people they have known for a long time. Indeed, a large network of contacts is an asset in creating and developing a new business, both to tap into knowledge and to develop customers. A frequent handicap of many entrepreneurs is that they are too isolated.

Chinese business culture

It is generally considered that China's management system places too much emphasis on hierarchy. Confucianism is often, at least partly, felt to be the cause for this. In the business world, executives may be considered as 'emperors', probably more so in SOEs than in private firms. This results in excessive submission to the boss, much politics around the 'emperor', and poor communication as well as, often, a lack of succession plan.

On their side, Chinese managers jokingly warn their foreign colleagues that PRC stands for 'Patience, Relationship and Cash'. They will also tell them a few facts about doing business in China:

1. There are strong differences between the various regions. Even the so-called coastal regions have their differences: Bohai around Beijing, the Pearl River Delta region close to Hong Kong and the Yangzi River Delta around Shanghai. On the other hand, people from Hong Kong and Macao as well as Taiwanese Chinese are basically treated as 'outsiders'.
2. Communication: in spite of their appetite for electronic communication, the Chinese prefer face-to-face meetings. In business meetings, they rarely go straight to the point. This is known as *TaiChi* talk.
3. Hierarchy is key: harmony is maintained, in part, by Chinese employees following the opinions and practices of the boss.
4. Chinese managers do not plan ahead.
5. Chinese 'China time': time may settle many things.
6. Bureaucracy is endemic.

7. Remain flexible: there is no need to make appointments much more than a few days in advance.

These 'Chinese ways' are indeed grounded in the country's culture. Elements of this are given below.

Elements of China's culture

This section intends to give a few elements of some of the characteristics of China's culture. This is done with the acute feeling that it involves sweeping generalisations and refer the reader to the bibliography where several authors have covered this subject more exhaustively. We will highlight only a couple of aspects here.

Chinese manifest a great pride for their history. The latter, it is sometimes said, is composed of '7,000 years of culture, 5,000 years of civilisation and 100 years of humiliation'. Traditionally, government officials were the most important citizens and successful individuals would donate large sums of money in order to secure government jobs.

In China, philosophy provides principles guiding the everyday life. It comes from three main sources: Confucianism, Taoism and Buddhism.

The all important philosophy from Confucius, born in 551 BC, puts a high emphasis on people as well as moral and social order. Its key elements are:

- *Li* is the ritual, understanding the etiquette.
- *Ren* is love.
- *Shu* means: 'do not do to others what you would not like have done unto you'.
- *Zhong* is loyalty.
- *Yi* means friendship.
- *Xiao* is filial piety.

Tao means 'the path'. It was developed by Lao Zi pretty much at the same time as Confucianism. It emphasises the harmony between man and nature and, more passive than the latter, it stipulates that a person may well find contentment by simply being one with nature. Both philosophies, however, celebrate *zhong yong* (i.e. harmony between nature and society).

Introduced to China in the first century AD, Buddhism has had a deep impact on Chinese culture, starting soon after the collapse of the Han dynasty in 220 BC. Compassion and wisdom are to be cultivated and the individuals are 'enlightened' through being in the present.

In the twelfth century, the Emperor Xiaozong suggested this way of summarising the Chinese world:

> 'Confucianism for the country, Taoism for the body and Buddhism for the soul'.

It is often said that the Chinese use symbolic and conceptual thinking (versus analytical in the West), looking at the 'big picture', before proceeding to see the details.

Another core concept in China is the law of opposites, the *yin* and the *yang*. According to this, the world is composed of two opposites, yet harmonious, forces. This approach is often interpreted to explain the way the Chinese envisage business transaction or the ability of managers to bring the best out of people working for them.

As Chinese companies become global, their managers and headquarters have to learn how to function outside China. Westerners have made many mistakes in doing business in China. Non-Chinese business partners need to establish open and abundant communication, while they must be open to Chinese ways and learn from them. The key is to retain the curiosity and tolerance of somebody else's point of view.

Young Chinese managers (below the age of 30) appear to be very different from their predecessors: quick, focused on the markets, internet-savvy, short-term orientated and with a relatively low tolerance to 'tradition'. At the risk of somewhat losing contact with their 'culture', young, urban Chinese are more international and global in their outlook. This, combined with a keen entrepreneurial spirit, is very positive for innovation.

---

[1] The report can be downloaded from www.oecd.org.

[2] 'Chinese returnees: Driving Force of Chinese Innovation'. (2014) Dr. Wang Huiyan and Xu Suining. See http://en.ccg.org.cn/Research/view.aspx?Id=764.

[3] See www.haigui001.com.

[4] Chinese Service Centre for Scholarly Exchange, Ministry of Education.

Chapter 7

The contribution of non-Chinese innovators

Foreign advanced technology has always been part of China's plan to modernise its society and help the country develop economically. What better way to avail yourself of this know how than having foreign companies bring the latest technology to China, together with the corresponding managerial know how to develop Chinese engineers in the arcane art of innovation? This is the underlying mechanism of how China attracted foreign companies to set up R&D centres in Shanghai, Beijing and many other Chinese cities, in exchange for 'access to market' – an access closely regulated and controlled by Chinese parties, as we have seen in Chapter 5.

A rapidly growing R&D presence

It all started with China's opening up in the wake of Mao Zedong's death in 1976, but it really took off only after Deng Xiaoping's famed 'visit to the South' in 1992, when foreign direct investment policies in China were developed and made more welcoming to foreign companies. The first point of attraction of foreigners was Shenzhen, right across the border from Hong Kong.

Back then, China's main advantage was low cost, both in terms of low salaries and low infrastructure costs. This attracted foreign manufacturing and production. At the time, the market in China was quite small: at approximately $500 billion at that time, China's GDP was about 20% smaller than Spain, with an average GDP of only about $500 per person. Growth rates, however, were impressive, and it was clear that China had an appetite for anything new and superior.

Large multinational companies came first, such as Philips, Motorola and GE, to establish product development units in the early 1990s.

China's potential as a future market, as well as a future source of talent, further attracted foreigners to set up research and technology offices in the mid-1990s. Between 1995 and 1998, IBM, Motorola, Intel and Microsoft started their research centres. With such reputable companies extending their trust to China, others soon followed suit.

For many, the market became the main driver, but co-location of process R&D with production facilities was also important. China's S&T Statistical Office and the MOST ministry have released regular updates on foreign R&D investment to draw attention to the benefits of doing R&D there. By 2000 it was estimated that there were 50 foreign R&D centres in China, 400 by the end of 2002, 750 by 2006, and 1,500 in 2012. Indeed, the size of this R&D presence varies greatly. Certain 'R&D units' are merely watching posts, with just a few staff members. Others count hundreds of engineers and large facilities. Philips, for example, has 2,300 staff in R&D, up from 650 in 2005. Sony has 2,000 staff in R&D in various locations in China. ST Microelectronics has more than 200 engineers in R&D, and the Dutch company DSM, which came in 2011, now has 120 R&D staff.

By 2015, most technology-intensive multinationals had R&D in China. Within 20 years, China has become the second largest host of foreign

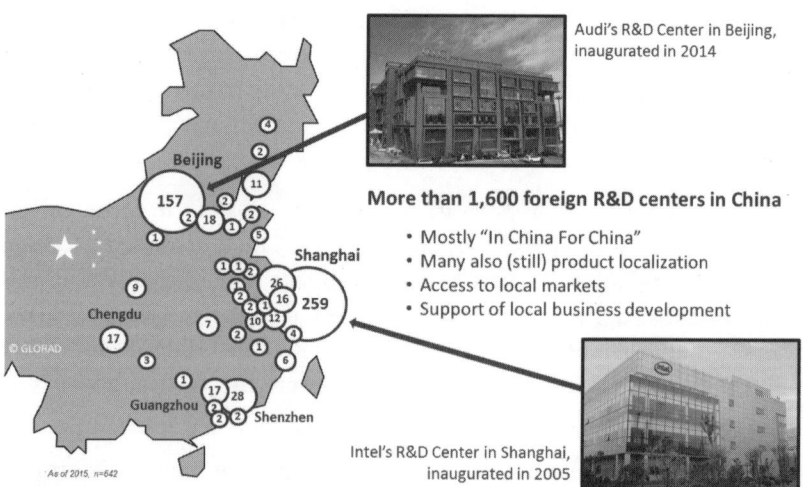

Audi's R&D Center in Beijing, inaugurated in 2014

More than 1,600 foreign R&D centers in China

- Mostly "In China For China"
- Many also (still) product localization
- Access to local markets
- Support of local business development

Intel's R&D Center in Shanghai, inaugurated in 2005

Figure 7.1 Location of foreign R&D Centers in China.

R&D in the world, after the United States (unless you consider the European Union to be a single entity, which would bump China to third place). China seems to be a firm component in the global landscape of multinational R&D.

Faurecia

Faurecia is a French supplier of vehicle interiors such as car seats and instrument panels. It has 97,000 employees and invests 5.4% of sales in R&D. A large part of its profits come from China, now the largest automobile market in the world, where both foreign and domestic companies serve increasingly mobile customers. The demand for innovative interior arrangements is very large. Faurecia has a growing R&D presence in China, with three centres alone in Shanghai. As an example, one of these employs 170 people (including only five expatriates) in developing emissions control systems for local manufacturers. A new technical centre was opened in 2013 which will eventually house 800 engineers. It includes five labs (testing lab, acoustics lab, airbag lab, complete seat lab and quality inspection lab), an industrial design facility, two showrooms and various prototype workshops, including the laser welding workshop. Faurecia already places great emphasis on having all of its R&D, including its China R&D, coordinated globally. China will soon produce cars that are sold worldwide, or buys itself into truly multinational car companies as we have seen with Geely and Volvo. Its impact is increasingly global.

The ambivalent attraction of China

It is difficult to resist the attraction of a huge and fast-growing market, especially if success is based on making products simpler rather than more sophisticated, and customers need any functionality they can get. Such were the early times in China, and in many cases they still are, particularly in the central regions away from the more developed coastal provinces of China. Innovation focuses on product reengineering and localisation, also known as 'de-featuring' when features are removed that are only rarely used but increase costs and thus affect the final price tag. Domestic Chinese firms have upgraded their products to compete with foreign companies in what is perhaps the toughest battlefield in China these days: the 'good enough' market or, as it is also known, the 'just right' market.

Foreign companies often have the upper hand in terms of technology but they struggle to make products that can compete with those made by Chinese companies. As local customers become increasingly demanding they buy foreign products, but Chinese companies also become better and thus what seemed to be a walk-over for foreign high-tech firms in China has really turned into one of the most performance-orientated competitive arenas in the world.

The battle over the 'just right' market

China is not just 'one' market. From a price point of view, there are often four co-existing markets:

1. The low-cost market: Here, consumers received undifferentiated service and quality for the lowest available price. Examples are many commodity products such as clothes, food, and toys.
2. The Western quality market: This consists, as the name says, of imported or made-for-export products sold in China mostly to high-income customers. Examples are expat food items, foreign medicines, Western automobiles, etc.
3. The 'world's-best' market: This consists of products in which China is the trendsetter or has aims to assume a leading role. Examples are battery technology, solar panels, and some super-fast train technologies.
4. The 'good enough' or 'just right' market: This consists of products of acceptable quality at a very reasonable price (and underlying cost). These are 'no-bells-and-whistles' products (i.e. those that do 80% of the job at 20% of the cost).

The 'just right' market is highly contested by Chinese companies with improving technology skills and their intent to expand profit margins, and foreign companies trying to defeature sophisticated Western products to make them appeal to a much wider local audience at a more affordable price range. These products may not sell well in their Western home countries, but they are 'just right' for the developing middle class in China.

Anyone who has ever travelled to China, Vietnam or Thailand will be aware of the VCD shops along the busy touristic streets in downtown areas. VCDs are video-compact discs, compressing movie information onto a medium which was originally designed for audio only. Not that VCD technology was not considered in the West, it was just bypassed in favour of the then-developing new DVD standard. In China, however, in the 1990s, VCD technology was 'just right': A cheap and mastered technology that was able to bring new benefits to a consumer base who wanted movies at a low cost and was willing to accept low quality and low usability (most movies were too big for one CD so the user had to swap VCDs mid-way through).

Foreign companies who are trying to find the right local entry point, the mid-end mass market, often struggle with redesigning their products for local needs. It is against engineering nature, perhaps human nature, to strip away features and functionality from a product that you worked hard to make possible in the first place. Local R&D teams are thus better suited to redesign for local needs, specific to China, using Chinese-specific materials, functions and handling, and building them for business models that work in China. Siemens, which has more than 20 years of experience with R&D in China, has coined this approach 'SMART innovation': Simple, Maintenance-friendly, Affordable, Reliable, and Timely-to-market. De-featured ultrasound machines for rural clinics in Western China are one result of Siemens's SMART innovation.

In order to attract more foreign R&D into China, the government in Beijing has repeatedly improved investment policies that provide foreign companies with tax breaks, tax holidays, and infrastructure when they set up companies with an R&D licence. Local municipal and provincial governments are more interested in large-scale manufacturing plants that employ thousands of people rather than a few hundred scientists, maximising tax revenue. The national policy is aimed at having foreign companies hire Chinese engineers and train them in technologies that are not part of textbook knowledge. This often takes years, as this knowledge is notoriously difficult to share and absorb. Once embodied, these R&D engineers can make significant contributions to the company, as part of local product development teams or also in global projects.

In China's hyper-competitive environment, young, well-trained engineers with a few years of relevant work experience in a foreign company, find it easy to get jobs with other companies, foreign or Chinese. As a result, staff turnover is high and damaging to those R&D-investing companies. Turnover rates of 25% are common, as compared to less than 10% in more mature markets. Not only do firms not get the full potential from their own R&D staff; they also see core technical knowledge leak away to other competitors. In countries where the institutions for intellectual property protection are well developed, this is less of an issue (albeit it is never eliminated). In China, especially in areas where the reach of the law isn't as strong as it could be, relatively poor IP protection undermines the competitiveness of companies, and their interest to invest in R&D. This applies to Chinese firms as much as it does to foreign companies.

Protecting intellectual property (IP) through R&D in China

China has a bad reputation as far as IP protection is concerned. As seen in Chapter 5, the origin of China's IP law only dates back to 1985. It is mostly the enforcement of IP rights that is substandard. This affects foreign as well as Chinese companies. Copycats copy only successful products, irrespective of who owns them or their underlying patents. The vast majority of patent infringement cases dealt with in Chinese courts are now Chinese companies suing Chinese copycats.

How can you protect yourself from copying? Unfortunately, keeping R&D outside China may not help; it may only delay you from learning that your product was copied in China, too. Having an R&D presence in China often helps to make sure which products are authentic and which have been copied, and in developing new local ways to protect technology effectively.

The first basic rule in protecting IP in China is to do everything as you would do it elsewhere. This means filing patents in China when they are part of products to be sold locally, registering trademarks in China, and keeping trade secrets. Perform store checks, monitor competitors at product fairs, scan the Internet for products similar to yours, and visit companies (including suppliers and customers) to find out more about how your product is being used (or misused). When you find copied products, set an example by destroying them and finding the infringing party. This is usually difficult and often not immediately helpful, as penalties are small and much of the detective work has to be done alone, often against the interest of local companies and their stakeholders. The best way is to educate everybody up and down the value chain about the detrimental effects of low-cost copying on everyday consumers – most of them Chinese, perhaps relatives of your employees! – if they operate unsafe machinery or consume hazardous products. Managing access to various types of data is restricting unwanted abuse of technical information. This starts with limiting product information on websites to user access controls in corporate intranets.

Can R&D help against IP theft? Of course it can, and if you have R&D on the ground in China, you can use the very same creativity that infringed on your technology to find new ways to help you protect it. One of the first defensive inventions was the introduction of hard-to-copy identification tags, but counterfeiters soon learned how to copy those! Black-boxing is another approach to keeping technology safe. The technology is encased in such a secure way that unlocking it basically means destroying it so completely that reverse engineering is no longer possible. Or design 'soft' features that are easy to notice but hard to copy (e.g. watermarks in currency bills, or a touch-and-feel (e.g., in automobiles)) that only those who truly understand and master the process can produce. Products that require a substantial amount of capital investment are usually safe, as the risks of detection before generating a return are too high for the infringer. Also, products that can be designed with a difficult-to-copy synergy feature (e.g. intelligent network services from an overseas base/router) will allow local counterfeiters to manufacture the hardware only, lacking the added-value from the out-of-China system on which the real consumer benefits are based. Local copycats also don't tend to be too advanced scientifically (although their level of know how is getting quite impressive), so anything that is very advanced or requires a unique skill set is difficult for them to reproduce. Designing R&D workflow in such a way that no single team has sufficient knowledge to replicate the entire product system ('work breakdown structure') is at least partially supported already by project management software.

It is important to remember that it is not R&D alone that prevents IP infringement in China. IP protection involves every function, some direct such as legal and sourcing; some less direct such as HR and marketing. It is also dependent on the proper signals and values communicated through strategy and top management, and the actions and demonstrations of leaders in daily operations. Also, foreign companies sometimes have cost advantages over the smaller counterfeiter, who may run batch sizes geared towards local consumption, not for global markets like the foreign multinational. Leveraging these cost advantages against local counterfeiters is an unpleasant bleeding-out experience for the foreign company, but it is part of the cost of doing business in China.

As seen in Chapter 6, approximately 6 to 7 million students graduate in China every year; one-third of them are engineers. They are young, they have a thirst for new things, and they are naturally very curious. In many ways, they would be the ideal R&D staff for any company. They also are the product of an education system that rewards memorisation and replication over innovation and teamwork. These shortcomings require strong training and coaching before local Chinese engineers can integrate seamlessly into the global innovation processes. In their early years starting out as R&D engineers, even though they are quick learners, they tend to be less productive and less a source of differentiation for foreign firms than they were initially hoping.

Things are constantly changing in China, and the latest generation of Chinese graduates is more international, more open-minded, better at speaking English and better prepared for doing R&D in foreign companies. However, R&D also becomes more expensive every year. Roughly only half of R&D costs are salaries (on average, across all industries, according to a recent study), and the rest are:

- Expenditures for facilities, offices, and infrastructure: These costs are often higher in China than in the West, especially in the overpopulated and dense cities of Shanghai, Beijing and Shenzhen. The only alternative for foreign firms is to relocate R&D to the suburbs, where the infrastructure is less convenient, or to second and third-tier cities, which local Chinese engineers find less attractive to work and live in.

- Costs for IT and telecommunication: Hardware costs are surprisingly high in China, given that most of the devices and equipment are actually manufactured there. It is often cheaper to purchase laptops, for instance, in the US. IT services may be less expensive, but they are also less transparent and reliable. Communications such as teleconferencing, e-mail, and the Internet are as cheap as you like, on paper, but the real issue is that bandwidth limitations within China make it very impractical to rely on public services. It costs money to set up private communications and this may still not overcome fundamental bandwidth problems.
- Costs of travel: Local travel is not as expensive as in the West, but international travel is. In any case, it is less convenient across China, with plenty of delays and challenges along the road. Beijing and Shanghai have airports that have among the highest rates of flight delays worldwide. Local traffic in cities is extremely cumbersome. Although local governments invest a lot in building public transportation, it is completely consumed and used to capacity almost immediately after being opened.

All of these problems mean that the advantage of a lower salary is quickly compensated for; the real overall cost advantage is small (15–25% at most) and vanishes the longer you are in China, due to increasing compensation packages for your more loyal employees.

Hiring the right R&D engineers

Despite having a huge pool of new engineering graduates to hire from, finding the right R&D engineers in China is actually not easy. For every job posting, foreign companies are inundated with thousands of nearly identical CVs. Students don't tend to get distracted from their studies by doing extra internships, taking part-time jobs for money, or other extracurricular activities that build differentiation into Western CV profiles. So Western companies rely on Chinese university rankings and recruit from the top schools. Their students have often taken courses on how to manufacture CVs (mostly because their innate CV writing skills are quite poor) so that they pass through the first reject gate at corporate HR departments, placing the responsibility back on the hiring R&D director.

Ideally, you want to talk to every new graduate with a great CV. But there is just no time to interview 100 seemingly equally qualified candidates for even just 30 minutes each. So what do we do? Two successful practices have emerged from two decades of R&D hiring practice: Background checks and automated testing.

Experience shows that it is necessary to check CVs for accuracy. Taken out of sequence, and with different HR staff screening candidates, sometimes identical CVs (identical in every detail except the name and phone number of the candidate) make it into the second round. Calling professors or former supervisors for references is necessary; verifying their authenticity is important, too. Eliminate candidates that don't check out. Before inviting someone for a face-to-face meeting, conduct a telephone interview. If they don't seem to have experience in the skills you seek, eliminate them – you will still have a few dozen candidates to choose from. Then add an automated round of skills testing. Think of it as an exam of their promised skills. This step can be done without getting the senior hiring management involved. Invite the candidates to a one-hour session, and have them solve a few representative mini-problems on site. Again, if they don't have the skills or speed you need, eliminate. Now you are ready for a personal interview, but make sure that when you talk to them, you have your team members interview them, too, to avoid only hiring those who happen to speak English well. Local team members are also better at sensing whether a candidate has the necessary cultural team fit. Even if everybody now agrees that the candidate would make a good addition to the team, it's not over. Make sure you introduce him to the company, what your purpose is here in China, and how he could fit in. Walk him through the terms of the contract, address open questions, and sound out potential areas of discontent. You want to hire smart engineers, but you also don't want them to harbor second thoughts and just wait for an opportunity or excuse to leave.

On-campus recruiting works in China as well as it works in the rest of the world. However, Chinese students are more brand-orientated than most. A household name such as Google, Microsoft or IBM will attract thousands, while smaller companies or companies in the business-to-business industries are less popular. Especially at top-ranked universities, students are wooed by dozens of attractive Western companies. Going to second-tier universities or to universities that are in central China but perhaps with a great reputation in your specific line often pays off. Another trick employed by some foreign R&D directors is to still go to top-ranked universities, but deliberately seek out candidates that graduated in the lower half of their class. While these individuals were clearly good enough to make it into an elite education programme, they may have spent their time at university learning all sorts of things, including developing their creative skills, rather than focusing on passing exams based on rote memorisation.

To get to qualified candidates more quickly, companies often institute referral policies that encourage employees to recommend fellow engineers. Such candidates often have a much better cultural fit than those who apply without any prior connection to the company; the referring employee has an interest to make sure that his colleague joins the firm and does well, too, especially if referral bonuses are paid out only after a grace or assessment period. Referrals are active network hunting; searching through professional networks such as LinkedIn and Xing, but especially through Chinese websites such as Ushi, JingBei, QQ or Weibo are great passive tools to identify good candidates. There is a great deal of network/platform-based communication among engineers from many different Chinese and Western companies which happens exclusively in Chinese and sometimes in 'coded' Chinese (to avoid censorship). Keeping an eye on such platforms is useful not just for recruiting!

Getting a university graduate up to speed (read: 'make him a net contributor to the company over the duration of his tenure with you') takes quite a bit of time: up to a year for the more menial repetitive R&D tasks, but easily many years in disciplines characterised by a great amount of tacit knowledge. If they quit before they become a net contributor to the firm, the company has essentially just subsidised China's technical education. Hiring a mid-level career engineer helps to shorten the time needed to make him a net contributor. Of course, mid-level engineers don't come cheap, and they command salaries equal to those paid in the West. Turning junior engineers into more senior ones is the challenge many foreign (and Chinese) companies are trying to get to grips with. Mid-career engineers tend to value job stability a bit more than before, as they now have family responsibilities, and they have also learned that super-fast promotions can get you into situations that are career-damaging rather than building.

One of the least tapped sources of engineering talent in China, and indeed over most of Eastern Asia, are women. Forty per cent of all engineering graduates are women, and they often have disadvantages in building careers the same way their male counterparts do. Female engineers tend to be more loyal to their employer; they value a good work climate over cut-throat competition and advancement at the expense of a social life. Such qualities are very important in R&D, where trust and openness among colleagues is crucial in cultivating an innovation-orientated environment.

Making innovation happen in China is hard

Managing R&D and innovation in a foreign country is difficult anywhere – it is particularly hard in China. Even if you speak Chinese, and perhaps grew up in China, there are several major headaches that will keep you awake during the arduous process of setting up and growing an effective R&D centre.

The human factor

The first concern is always people. As mentioned in Chapter 6, these days concerns with human talent constitute a more acute preoccupation than intellectual property. How do you manage your engineers, how do you train and coach them, and how do you lead them to make the most of their potential? How do you keep staff turnover below 20% – ideally below 10%? How do you win their trust so that they stay loyal to you and the company? There are no easy solutions; they will require full-time dedication, not just at a professional level and not just during office hours, but also from you as an individual leader and genuinely caring boss concerned with the wellbeing and welfare of your subordinates. One of the first lessons an incoming R&D director learns in China is that he will not just direct: he will also be a coach, a manager, a dictatorial boss, a forgiving father, a technical expert, a late-night-shift intern, a teacher. This is a full-time job. (More on this is to be found in Chapter 6.)

In fact, most of the critical work in leading people seems to be done after office hours. It will require you to get to know every one of your R&D employees on a personal level. Some Western directors are uncomfortable with this, and it may even be against company culture or regulations to get so involved with subordinate ranks. However, this is Chinese leadership style. If you don't master it, your key engineers will eventually find organisations where they feel more appreciated and welcome, and you will need to find new top performers. One of the problems in R&D in China is that you often finish important (i.e. strategic) R&D projects with different people than you started with, because of high staff turnover. This is not good for meeting quality, cost and time milestones, let alone innovativeness of your solutions.

Retaining one's technical position

The second headache is technology leakage, IP control, and retaining know how. This goes beyond managing your company's patent portfolio in China. Irrespective of how well you believe your technology is protected, there is always the possibility of key know how bearers walking away and taking this technology with them. It doesn't matter that you own your patents and that your former employees have broken the law. Of course, according to Chinese law what they are doing is illegal and you have the right to cease and desist their operations. In reality, the obligation is on the patent holder to collect sufficient evidence that infringements have actually occurred, and that they are significant enough to warrant action. The police will not act on your suspicion, hunch or fear in isolation. The evidence will have to be both concrete and actionable. Unfortunately, this often means taking much of the investigation into your own hands, which is borderline even in Western legal contexts. It also means that the presumed perpetrators quickly realise that something is coming their way, and they will cover their tracks.

Recent cases in China have backfired and brought major multinationals to their knees. Even if you have an iron-clad case against a former employee, are you really willing to risk the bad publicity when the press will bemoan how this 'rich foreign company chases down a poor

Chinese engineer to squeeze him out of his livelihood?' The damage can be irreparable, and significantly higher than the leakage of the technology itself. Managing IP in China requires you to be aware of these sensitivities, have a firm moral ground, and manage expectations multi-laterally both inside and outside your own organisation.

Making collaborations work

Managing stakeholders outside the R&D centre is not limited to IP issues. It extends to establishing meaningful two-way exchanges with other research institutes, joint product development with key suppliers, involving lead customers and clients in anticipating new market trends, working with technology park officials to secure preferential access and real estate for your R&D operations, even getting enough face time with local mayors and provincial decision makers. Those are just the obvious stakeholders for your R&D organisation in China! Often, the real movers and shakers operate behind the scenes; it takes a lot of time and effort to understand who they are and how they benefit from seeing you succeed. Let us also not forget your own country's organisation, with whom you need to sync up over product and market plans, and how you can support those. Given the high growth, and often quickly changing demands, this is no easy task. They will want you to focus on their problems, their growth opportunities, since you are, after all, located in their market and on their home turf. They probably pay a substantial share of your expenses, so they have a strong argument for why you should care primarily about their innovation needs.

Global headquarters are so far away

Headquarters are so often the greatest obstacle to business development. They seem to know everything so much better, exporting totally irrelevant practices to the various corners of the globe. In so many cases, business in China took off only when the 'centre' gave real autonomy to the 'periphery'.

It is a very similar situation when it comes to innovation. Either the headquarters hold you back on your natural progression of innovative capability. In this case, the local R&D outfit has a carefully defined role to play in the global concert of R&D contributors. The urge to expand the innovation portfolio by assuming greater local responsibility is disturbing this picture. Or, the 'centre' wants to see real 'bang for their buck' and demands that innovation productivity increases quickly. After all, they have supported it for several years, and all that seems to have happened is the hiring and training of more people, keeping average productivity low.

On top of that, salaries and other costs are increasing in China, reducing year-on-year productivity. In fact, cost increases in China are easily at 15% per year, which means one would need to increase productivity of R&D staff – which suffers from high staff turnover and strong distraction to train newcomers – by 15% across the board just to maintain productivity levels at the level of the year before. Of course, headquarters are asking for real productivity increases, which leaves you, the R&D manager in China, with the daunting task of telling your supervisors back at headquarters what you are actually doing in China, what is really going on there, and what you would need from them in order to build an effective R&D organisation that contributes to both local and global needs.

Samsung in China

The head of Samsung China, Ho Moon Kang, recently declared: 'In China, we want to create business models and products designed to serve the Chinese market first, and eventually export them to the rest of the world.'

In order to achieve this, Samsung has invested heavily in R&D in China, with plans to have 5,500 R&D staff in China by 2016, representing more than $300 million in investment.

Samsung has 91,000 employees in the Greater China region alone. With revenues of more than $60 billion in 2014, the lion's share (80%) came from mainland China, where Samsung has invested $10 billion so far.

Currently, in Greater China, 22 Samsung companies manage 144 units and offices. Through its various subsidiaries, such as Samsung Electronics, Samsung Semiconductors, or SDS, it expands R&D simultaneously in different industries in key target cities. So far, Samsung has also set up seven research facilities, providing a complete chain from conception to manufacturing and sales. For instance, Samsung made an additional $500 million investment in 2013 that included a new R&D centre on top of a $7 billion investment for a new semiconductor plant in Xi'An. Apart from semiconductor technology, this new R&D centre also works on smartphone and TV technologies. Covering more than 4,000m^2 of lab space, the centre works closely with local universities, many of which have an excellent reputation for their research on information and communications technology. With a target staff size of 200 engineers, it is also tasked to serve as an advance base for further development of the Western inland region of China.

Is it worth it? – The foreigner's perspective

China is not an easy place to do R&D, but it may very well be worth the trouble. The value of doing R&D in China contributes to global business at many different levels, and ideally a multinational firm is able to leverage all of them.

Some of the foreign R&D centres have grown quite impressively in size, scale and scope. Motorola had, at one point, 19 R&D sites in China; its R&D centres having developed finger-writing technology for smart phones and the 'Ming,' a very popular phone in China. Before its acquisition by Microsoft, Nokia had moved part of its core R&D for handsets to its Beijing R&D centre. It is estimated that 5% of R&D engineers in China designed 40% of Nokia phones worldwide. Microsoft itself was an early pioneer. Its research centre was set up in 1998, and by now is as productive in terms of papers, patents as any other Microsoft R&D site. With increasing skills and capability, companies entrust their Chinese organisations with more responsibility. ABB, for instance, moved not only its worldwide robotics R&D centre but also the business unit's headquarters to Shanghai, as early as 2005.

A further challenge is for non-Chinese small and medium-sized enterprises (SMEs) to participate not only in the Chinese market, but also in its growing technical development. SMEs often follow a lead customer (i.e. a multinational firm that they supply components to) and which demands that its key suppliers are present at its major manufacturing and R&D locations to ensure complete design and development capabilities. But what constitutes a small extension of a few people for a MNC of 100,000 people may be a strategic investment for a SME of 1,000 employees. Beyond the capital, they also need advice and managerial know how, as they may not have sufficient international experience, especially in China, within their management. They also don't have a brand reputation that would help them attract top job candidates, and they don't have the political weight to throw around (in terms of potential jobs created or lost) in case of unexpected administrative or legal problems. It's not easy for a foreign SME in China. Hence, several programmes have been set up to assist SMEs. Some have been created by the home countries of these SMEs (e.g. the German Centres in Shanghai and Beijing) and some are set up by local Chinese institutions, such as China's Torch program, which often help to make connections between foreign and local companies and identify partners for co-development.

Patents granted to foreigners

At a fundamental level, foreign R&D in China produces technology. Every year, foreigners apply for more than 100,000 patents at China's State Intellectual Patent Office (SIPO). In 2014, 127,042 patents were applied for; 70,549 patents were granted to foreigners (compared to 162,680 patents granted to Chinese inventors). In total, foreigners owned 653,344 patents in China (about 42.1%) of all invention patents in China. The share is smaller for utility models and design patents, which are more the domain of Chinese inventors. In the US, the equivalent numbers were 578,802 patent applications, and 300,678 patents granted – of which 49% went to US inventors. As already discussed in Chapter 3, China already produces roughly 80% of the number of invention patents in the USA, while foreigners' inventions in China account for about a quarter of what the US invents at home. These inventions are owned by foreign companies and find their way into global products and technologies.

Evolution of the non-Chinese R&D in recent years

Even if R&D staff attrition is unusually high, many of them stay and become excellent scientists and engineers. Microsoft, IBM and others have stated that their China R&D centres are as productive as any elsewhere in the world, contributing to product development, technology and science in equal shares. As these individuals rise up in the hierarchy, they get overseas assignments where their know how makes a difference in local R&D work outside China. Many senior researchers in the West are of Chinese origin, having started their careers in those early R&D labs, and now act as ambassadors for their home country, or as conduits for ongoing R&D collaboration between China and overseas R&D bases.

Foreign R&D in China often serves a national policy purpose. In some industries, local content rules require foreign companies to conduct a certain amount of R&D in China, usually as a percentage of local value-added in products sold locally. Not doing R&D in China means not being able to sell in China, and even the most repetitive and menial R&D in China can still maintain a business presence in China, as long as its costs don't exceed local profits.

In the 1990s and early 2000s, much of the R&D effort was driven by exploiting cost advantages: low labour costs, subsidised infrastructure, low rent. At that time China's innovation performance was still very low, so most companies considered investing in R&D in China as a real option. China's role as a manufacturing centre also helped, as it was convenient to locate development teams close to production facilities, where they could work on process or product improvements, leverage the infrastructure and the administration overhead, serve as a training centre and assist with quality assurance. In some cases, especially if you came early, you would be able to hire the brightest of the brightest and put them to work on your long-term research.

China soon became attractive as a market in its own right. Following the classical evolution of the product life cycle, companies first sold old products to China, and then started to localise them by making inexpensive but market-specific adaptations. As China reached a certain size in terms of global revenue, China-specific products were included in the blueprints of the original product development, until, eventually, some products were designed and developed for China first or for exclusive sale in China. Once the local R&D team was competent and mature enough, this responsibility was shifted to China.

Currently, most foreign companies are 'in China for China', and are happy if they succeed at this mission. Their R&D centres serve an important purpose: to satisfy the needs and demands of a very large market in the world, and to provide the parent multinational company with a secure source of revenue. This is and should be the primary goal of any innovation unit. China's context-specific demands, user language interface requirements, relative geographic isolation, and its own rate make it easy for China-based R&D centres to focus on China first and only.

Created in China for the world

The real win for multinational companies is when their Chinese R&D centre starts to develop global products for the world. Bayer, of Germany, recently decided to make its Shanghai laboratories world leaders for the development of a number of specific products. In 2015, this company had an R&D unit staffed with 300 engineers, double of what it was three years before.

China-originated products are celebrated as the next 'big thing' in global innovation. Most multinational companies simply are not set up to recognise and leverage the so-called 'reverse innovations' from 'emerging' economies, even when they occur. The local R&D centre in Shanghai, Beijing, Chengdu or elsewhere in China may be perfectly capable of coming up with the next breakthrough technology, but internal resistance, the mismatch of local and global incentives, and the predominance of the biggest market suffocates many good ideas.

Reverse innovation from China

What is a reverse innovation? The most widely used definition covers products that were first sold in a 'developing country' and were later introduced to an industrialised country. It is not always clear what is a developing country and what is an industrialised country; countries may be both developing and industrialised depending on where you look. Until recently, except for the most advanced parts of the coastal region, China is still a developing country. What about products developed in China but only for advanced-country markets in the US or Europe?

Reverse innovation may be developed by local firms, so Chinese companies developing and innovating products for China first and selling them later elsewhere are also engaging in reverse innovation. This little excursion just serves to show that even what seems to be a simple definition easily becomes complex in the realm of constantly evolving socio-economic realities. There is, however, somewhat of an ethnocentric, not to say racist, view for the West to assume that south–north or east–west innovation is 'reverse', while 'proper' innovation is supposed to be the other way.

Besides GE's low-cost portable ultrasound machine, such innovations include Nokia's mass-market entry phones, or Carel's air-conditioning controllers invented in China and later sold worldwide. The origin of many reverse innovations are just smart (SMART in the Siemens terminology) applications of local product (re-) development; the parent company then realises that by a stroke of luck these new products also appeal to customers elsewhere. What sells in China might not sell in Denmark, not least because the two countries have a very different product-price-quality balance. Of course, they may sell a few in Denmark, as there is always an overlap of preferences in certain market segments in different countries. Still, it may not be justified to call these lucky wins 'innovations'. Given the differences in market preferences between China and most industrially advanced countries, most of these future innovations will come more directly from the significant patent and technology applied to Western needs, rather than Chinese products reapplied to Western customers.

Is it worth it? – The view from China

China is already on the winning side, and has been since the first R&D centre was opened there. Foreign firms provide an upfront investment in terms of training, technology transfer, and even locally developed technology that benefits Chinese engineers and Chinese science. Chinese employees mostly work on localisation innovation for Chinese markets or, better still, on China-specific new product development. There has been some concern that foreign companies hire away all the good engineers. Recent surveys suggest this is no longer the case. Many join local Chinese companies or government or state-owned enterprises, and their work presumably benefits innovation in these companies. Others start their own companies, leading to better and safer products, which is in everybody's interest. The longer foreign R&D is in China, the longer the Chinese innovation system benefits. The better the Chinese innovation system becomes, the more non-Chinese companies will benefit from it.

In recent years, in financial terms, R&D investments have represented roughly 2% of the foreign direct investments (FDI) into China. When it comes to innovation, multinationals bring much more than money. Managerial processes and know how, as well as an innovation development mindset are even more valuable contributions to the country.

In brief, many multinational firms have set up R&D activities since the mid-1990s. Most recently, in the period 2012–2015, a new wave of growth seems to have happened, based on the experience gained and the willingness to take R&D to the next level. Bayer, Dupont, General Electric (GE), are all increasing their R&D presence in China. As an example, GE now has 3,000 engineers in R&D in Shanghai and Chengdu, the latter in the area of medical equipment. Companies also continue to 'arrive' in China, such as the French leader in industrial gases, Air Liquide, which founded a Shanghai R&D unit in 2013.

One of the greatest opportunities in China is to be developed by pharmaceutical and life-science companies. China's population is aging, and the transition in lifestyles increases the prevalence and incidence of many diseases that pharmaceutical companies are researching or have products for in the market already. Furthermore, R&D may be done more efficiently in China, and clinical studies can be carried out at greater speed and for a fraction of the cost of elsewhere. This has prompted Pfizer, GSK, Sanofi, Novartis and Roche to make large investments in China. Swiss firm Roche, which operates in China under the name of Shanghai Roche Pharmaceuticals, has had an R&D centre in Shanghai since the mid-1990s and a dedicated research facility since 2004. In 2015 it announced an investment of close to $500 million in a new diagnostic manufacturing facility in the Suzhou Industrial Park.

For most multinational firms, the end goal is to develop products for China in China, but also, and increasingly, for the world. In the coming years, their daunting challenge is to integrate their innovation activities in China with their global network. This will be a difficult task, considering how difficult it is to have an effective circulation of ideas and knowledge in their current array of units in the Western world.

Multinational companies have contributed to China's know how in managing innovation. This includes knowing how to plan and execute multi-functional innovation projects, handling the tools and software to manage them, the timely sharing of information, the capacity to integrate various parts of the innovation puzzle as part of system integration, the crafting of projects relevant to both the firm's business and the global organisation, and much more. These contributions were in part made by foreign expats but to an even greater extent by Chinese overseas returnees. This know how will ease the transition of Chinese firms on the path from product adaptation to technology creation. It constitutes an important element in shaping the patterns of China becoming a global innovator. We now turn to these patterns, as they concern the government sector, on one hand, and companies, on the other hand, in the two following chapters.

Chapter 8

Becoming a global innovator – patterns in the public sector

In the previous chapters, we have described the actors of the innovation-led wealth-creation process, as well as the 'framework conditions', which help shape China's innovation scene. Now will be discussed the specific patterns of the policies and practices followed by these actors, which are specifically designed to propel China to become a global innovator. These patterns concern: (1) the governmental and administrative apparatus, discussed below, and (2) the companies carrying out innovation-intensive business activities. The latter will be discussed in Chapter 9.

A pro-innovation leadership

China has a uniquely strong, vertical and pragmatic government in tight control of the country, although the latter is gradually becoming less of a command economy, as we saw in Chapter 4. Its leadership is remarkably committed to innovation-led wealth creation. This orientation is similar to that of Japan and Korea, two Asian countries that lifted their economies from poor to wealthy within just a few decades and which, in some ways, serve as models for China's trajectory.

Early in his launching reform in 1978, Deng Xiaoping focused on science & technology. The State Science Commission convened 20,000 experts to develop a plan for science to serve as a key engine for the country's economic development. Ever since, science policy has been firmly in the hands of the top leadership.

The majority of China's top national leaders have engineering degrees or a technical background. President Xi is a chemical engineer, with a PhD in Law, also from Tsinghua. This helps appreciate the power of science and technology in fostering economic development. In the seventeenth Politbureau of the Standing Committee, eight out of nine members have an engineering degree. In the USA, only one cabinet member has a similar degree. Many Chinese leaders have studied at Tsinghua University, in Beijing. Japan's leaders normally graduated from the University of Tokyo's Law Department.

When it comes to the policies specific to foster innovation, the central government sets the priorities and defines the policies. Provincial governments are tasked to implement them. They do not always do this with optimal diligence and competence, sometimes blunting the edge of the Beijing directives. This is guided by provincial and local conditions, allowing for a positive degree of diversity. In their interpretation, local governments emphasise certain aspects of the national policy over others, occasionally adding complementary regulations in order to suit local needs.

By and large, local governments are highly sensitive to keeping and increasing employment. In this sense, they may give priority to keeping a large manufacturing plant going, rather than to promoting innovation. Even though the party and the country are very large, its governmental apparatus is often surprisingly diverse, flexible and adaptive. We look below at how innovation policies are developed and at the specific example of 'smart cities'.

Innovation policies

As in any other country, the process of developing policies in China is reasonably chaotic. There is no one clear path. Each ministry, as well as the party and the regional governments, have their advisers. Institutes offer valuable resources, but have to navigate among many actors, such as the Development Research Centre (DRC), which is at ministerial level and is tasked with formulating the basis of new innovation policies.

The municipal governments also represent a force. This is particularly the case of the powerful region of Shanghai, China's main business centre, with 23 millions inhabitants. In this world-city, for example, the municipal government has a Science and Technology Commission of the Shanghai Municipality. The latter sets the objectives and policies for Shanghai, liaises with other departments from the municipality and is involved in attracting firms, large and small, to the city, and in putting people in contact, in order to develop the science base of the Shanghai metropolis.

Strong local governments and/or powerful local leaders make a difference in places such as Wenzhou, Chengdu or Dalian. This is similar to what happens in other countries; the difference is that, in China, such leaders have the machinery of the disciplined Communist Party to implement actions. A notorious example was Bo Xilai, a former Minister of Trade, and son of a victim of the Cultural Revolution. While mayor of Dalian, he transformed the city. As governor of the north-eastern province of Liaoning (which is also the name of the first Chinese aircraft carrier), he strongly favoured economic growth. As party chief of the city-province of Chongqing, he waged a vigorous campaign in favour of economic development, social services, as well as anti-crime. In 2013, he was sentenced to life in prison on corruption charges.

Developing innovation programmes for 'smart cities'

In some instances, Beijing launches an initiative towards innovation and economic growth, but does not have the budget for it. One good example is the case for the programme of urbanisation launched in 2014. The plan is to add millions of people to the urban population of China by 2020. The people living in cities will then represent close to two-thirds of the country's total population. The cost of building such new cities is supported by the provincial and municipal governments. For one, the Shanghai municipality has decided not to be part of this project, perhaps not surprisingly, since this municipality already has one of the most developed city infrastructures in the country.

In March 2014, the new urbanisation policy was initiated by the powerful National Development and Reform Commission (NDRC), which published a plan to have 100 million people move from the countryside to cities by 2020. To comprehend the magnitude of this undertaking, this corresponds to the equivalent of building ten New York cities.

The new cities primarily concern ten sites, including Harbin, Hohhot/ Baotou, Taiyuan, Ningxia along the Yellow river; most of them are in central and Western China. Beijing feels obliged to have a leading role. Thus, the city of 'JingJinJi', the future Beijing metropolis will include

Tianjin and Heibei, totalling 130 million inhabitants. China is remarkably good and fast at building real estate and, hopefully, will do so in a human-centric way for these new cities.

The design, planning and building of such new cities offers enormous scope for innovation in providing an urban environment pleasant to live in, organising space, optimising its use, reducing the impact on the environment, providing effective management of energy and traffic, and also for the general administration of the city, healthcare services, minimising energy consumption, as well as providing effective information on municipal services, such as nurseries and schools, or available housing. Such urban zones, loaded with information technology and data, are often referred to as 'smart cities'.

In concert with the urbanisation programme, several smart city programmes have been launched by Beijing; these will benefit both existing cities and the new ones (e.g. the Ministry of Housing and Urban-Rural Development has 90 pilot projects in this area). The various national programmes on the Internet and 'informatisation', mentioned later in this chapter, including those from the MIIT, include components for smart cities. The Shanghai exposition in 2010 was used as a testing ground for several of these programmes. A 'China smart city conference' is to be held early September 2015, in Shenzhen.

It is anticipated that the business volume in the broadly defined area of smart cities will represent $6 billion in 2020 for China alone. This area is natural for China, because of its very large and dense urban population and as a result of the goal of making China an 'internet country'. Let us hope that such a great opportunity will not be wasted, so that these new cities will be built as good places for people to live, something which urbane Europe has uniquely managed to achieve in the world.

For China's pragmatic government, a pattern of experimentation

In spite of the heritage from the Soviet Era's ways of organising and managing the state, the Chinese government is pragmatic and willing to experiment. A recent example of this pattern of 'trial and error' is the creation of Shanghai's 'Pilot Free Trade Zone' covering an area of 120 km². The initiative of China's Prime Minister was announced on 29 September 2013. At that time, the specifics of this sweeping initiative were not at all clear either to firms or the public at large. It was not specified what would be the rules of the game in the zone, and how the bureaucracy would be made lighter for enterprises, financial services and trade. At some point, the media even mentioned the unrealistic possibility that the Chinese currency would be freely exchanged in the zone. Much hype surrounded this initiative, as well as much uncertainty. The government asked for patience, announcing that the policies and rules for the zone would be finalised before 2015–2016. This pilot zone is for reform, not just for trade and financial services. It is rare for any government to announce such a bold initiative (potentially, the zone could be one similar to Hong Kong), while being ready to experiment as a way to test reforms regarding financial services and the bureaucracy of the administration, in particular.

So far, the long government list of activities, controlled or banned, in the zone has somewhat damped down the excitement. A clear result of the announcement of the creation of the zone has been a rush for companies to get established in it. This has resulted in a sharp increase in the price of real estate in that district.

At the local level also, the approach is often very pragmatic. In the mid-2000s, Qingdao (where the celebrated beer is produced, following the nineteenth century German influence in the region) decided to stimulate the usage of electric motorcycles and penalise those powered by combustion engine. A company was encouraged to produce these in large numbers. The government then offered free electric recharging of batteries to its employees, further spreading the adoption. Today, China is the largest producer of such bikes, with a production close to 25 million units per year. Their ballet in the Chinese cities is remarkable, as they move silently in massive numbers.

Programmes and policies to stimulate innovation

These essentially concern technology-intensive innovation and R&D investments. One basic instrument of China's policy is the plan for the

National Science & Technology Development for the twelfth five-year period (i.e. 2011–2015). This document was prepared by the Chinese Ministry of Technology (MOST) and the National Development and Reform Commission (NDRC), already mentioned earlier in this chapter, the Ministry of Finance (MoF), the Ministry of Education (MoE), the Chinese Academy of Sciences (CAS), the Chinese Academy of Engineering (CAE) and other bodies.

The CAS 'incubating' the Legend company

The Chinese Academy of Sciences acted as an incubator for the well-known Lenovo computer company. In 1984, eleven scientists at the Institute of Computing Technology of the Chinese Academy of Sciences (CAS) started the Legend Group under the auspices of the CAS. The first big success was their solution to the 'Chinese Character Processing Project', which was launched by the government to solve a problem of PC systems in transmitting Chinese characters onto screens. Then CAS invested RMB 200,000 into the company. Unlike their competitors, including giants such as Microsoft, Legend decided to set its focus on the development of new hardware instead of software. The result is known as the 'Legend Chinese-Character card', which provided tremendous advantages over Chinese software at that time. The company's innovative achievements were honored by the National Science-Technology Progress Award, the highest accolade for scientific developments. Despite this recognition and the financial support by CAS, Legend faced considerable bureaucratic barriers in its distribution business. In order to push sales of bundled foreign computers with their Legend card, the founders decided to leave the institute and move to Hong Kong, where they incorporated the Legend group.

In the following years the company experienced a rapid growth. The new name 'Lenovo' was adopted in 2003 – with the intention to indicate the roots on the one hand (Le-gend) and on the other hand the new innovative spirit of the company ('novo').

Another document is the Plan for Science and Technology Development 2006–2020. Soon after assuming power, President Xi ordered a mid-term evaluation of this programme.

It is sensible to be somewhat sceptical about the true importance of such documents. In the West, we have reason to be so when considering, for example, the bombastic statements of the USA 2014 State of the Union Address on a 'smarter national security approach', a 'strong and principled diplomacy' and 'vast improvements in the US education system'. The European Union (EU) is no different, with its March 2000 Lisbon declaration (or Lisbon agenda) on 'making the EU the most competitive and dynamic knowledge-based economy in the world'. Such utterances are devoid of serious and sustained political will and do more harm than good.

In the case of China's innovation/S&T, however, policies supporting them have been so consistently backed up by the government and the party that the general thrust of documents in this arena are somewhat more credible than in the west. In recent years, policy documents and media often refer to the need for China to develop 'indigenous innovation' (Zizhu Chuangxin) or home-grown innovation, as a way to move up the value chain of offerings and favour domestic demand as well. Indeed, partly as a result of the rapid rise in salaries, China often assembles components, produced in Vietnam or Indonesia, capturing a small share of the profit.

In addition to declaring seven 'strategic' emerging industries (SEI, see Table 8.1 below: energy conservation and environmental protection, next generation IT, biotechnology, high end manufacturing, new energy sources, new materials and new energy vehicles), the NDRC plan for 2011–2015 includes the following specific objectives:

- further improve the intellectual property (IP) system;
- increase R&D investments to 2.5% of gross domestic product (GDP) by 2020; and
- reduce China's reliance on foreign technology to below 30% by 2025.

Another policy document is often referred to outside China. It is the National Medium and Long-term Programme for Science and Technology Development (2006–2020) issued early in 2006. This ambitious long-term plan has the goal of making China a technological powerhouse by 2020. This comes with equally ambitious objectives for the investments in R&D, as well as the output in patents and publications. These metrics, and some of their limitations, have been discussed in Chapter 3. Several tools to implement an improved innovation performance include the following programmes:

- The Torch programme, launched in 1988 by MOST, has created 54 'science and technology industrial parks' in the country. It attempts at being decentralised and aims at incorporating private small and medium-sized enterprises (SME) into a positive momentum for innovation. It is claimed that 60,000 firms employ close to eight million employees in these parks. The first zone was Beijing's Zhongguancun, discussed below.
- The Torch programme also provides a seed investment fund called Innofund for small firms and more than 1,000 incubators.
- The National High Technology R&D Programme '863', founded in March 1986, funds, via the Chinese Academy of Science (CAS), mainly public research institutions and state owned enterprises (SOE). This programme provided the bulk of the financing of the Tianhe 2, the world's fastest computer. Quoting from the website of the Ministry for Science and Technology:

> In line with national objectives and market demands, the program addresses a number of cutting-edge high-tech issues of strategic importance and foresight during the 10th Five-year Plan period. They are:
>
> (1) Develop key technologies for the construction of China's information infrastructure.
>
> The 863 Program will focus on developing a number of key technologies in the next five to ten years and establish systems of significant value for application. It aims to accelerate the national socio-economic development, drive industrialization through informatization, and enable China to approach or catch up with international pioneers in selected fields by the year 2005.
>
> (2) Develop key biological, agricultural and pharmaceutical technologies to improve the welfare of the Chinese people.
>
> The 863 Program will concentrate on developing key technologies in agriculture, pharmaceuticals, and other related areas. It will enhance

the overall bio-technological R&D level and capacity by a significant margin.

(3) Master key new materials and advanced manufacturing technologies to boost industrial competitiveness.

- The National Key Laboratories Programme, started in 1994, aims to establish world class laboratories for basic and applied research, including collaborative work with firms.
- The '973' National Basic Research Programme, which was founded in March 1997, represents roughly 10% of the Science & Technology funds from the central government, for agriculture, energy, IT, materials and healthcare. In 2012, a budget of the equivalent of about $350 million allowed the funding of close to 200 projects. Their duration is no more than five years. European researchers can participate with Chinese colleagues.
- The Key Technologies Programme was founded in 1982. It constitutes the largest effort on scientific issues concerning China's economic development. It is managed by MOST and represents close to 20% of science & technology investments from the central government.

All the above programmes focus on technology-intensive innovations. Similar programmes are found in other countries, Japan in particular. The keywords of such programmes are often identical: nanotechnology, big data/analytics, smart cities, advanced materials, digital revolution. Although implementation and coordination are far from being flawless, the remarkable aspect of China's programmes is that they are energised by the relentless commitment to innovation-led growth on the part of the executive branch of the government.

As mentioned earlier, provinces adopt different ways to promote economic development through innovation-led growth. For example, while Beijing and Shanghai have both relied on strong governmental guidance, the Guangdong province (home to Guangzhou and Shenzhen) had a fairly hands off, unplanned approach, leaving it up to foreign investors to make the choices, resulting in manufacturing of IT and telecom equipment for export. The global telecom companies of Huawei and ZTE were founded in Shenzhen in the 1980s.

Table 8.1 The Seven Strategic Emerging Industries (SEI), according to Chapter 10 of the 12th Five-Year Plan (2011–2015) for the National Economic and Social Development of PR China.

Industry	Sub-industries
Energy conservation and environmental protection	Key technologies, equipment and products for efficient energy conservation; critical generic technology R&D for resource recycling; advanced environmental protection equipment and products; market-based energy conservation and environmental service system; waste and old product recycling and reuse system underpinned by advanced technology; clean coal and seawater utilisation
New-generation IT	Information network infrastructure, new generation mobile communication; new-generation internet; new generation of core equipment and intelligent terminals; three-network convergence; internet of things; cloud computing; IC; new displays; high-end software; high-end servers; software services, internet-based value-added services and other information services; intelligent infrastructure; digital virtualisation
Biological/biotech	Biotech medication, new vaccines and diagnostic agents, chemicals, modern Chinese medicine and other innovative medicine; biopharmaceuticals; medical devices, materials and other biomedical engineering products; bio-breeding; green bio-products for agricultural use; bio-agriculture; bio-manufacturing; marine biological technologies and products
High-end equipment manufacturing	Main and branch line aircrafts, generic aircraft and other aviation equipment; space infrastructure; satellites and application thereof; passenger special lines, urban railway transportation and other rail traffic equipment; marine engineering equipment; intelligent manufacturing equipment based on digital, flexible and system integration technologies
New energy	New-generation nuclear energy and advanced reactors; solar energy utilisation; photovoltaic and photo-thermal power generation; wind power technological equipment; intelligent power grids; biomass energy
New materials	Rare earth functional materials, high-performance membrane materials, special glass, functional ceramics, semiconductor luminous materials and other new functional materials; high-quality special steel, new types of alloys, engineering plastics and other advanced structural materials; carbon fibres, aramid fibres, ultrahigh molecular weight polyethylene fibres and other high-performance fibres and compound materials; nanometer, superconductive, intelligent materials and other common basic materials
New energy automobiles	Power batteries; driving motors; electronic control; plug-in hybrid electric vehicles; pure electric vehicles; fuel cell automobile technologies

Manufacturing plan

Another way to foster innovation is to stimulate China's massive and powerful manufacturing. In the Spring of 2015, the Ministry of Industry and Information Technology (MIIT) provided some information on the 'China Manufacturing 2025' plan, after approval by the State Council executive meeting in Beijing. The ten key industries due for a manufacturing upgrade are:

- new-generation information communications technology
- high-end computer numerical control machine tools and robots
- aerospace equipment
- marine engineering equipment and high-tech ships
- rail transportation equipment
- energy-saving and new energy automobiles
- electric power equipment
- new materials
- biological medicine and high-performance medical devices
- agricultural machinery and equipment.

The Plan also calls for five large projects, as follows:

- construction of national manufacturing innovation centres
- intelligent manufacturing
- industrial base capacity and quality enhancement
- green manufacturing
- high-end equipment innovation.

The plan will be implemented in line with the 'Internet Plus' strategy, emphasising the importance of integrating 'informatisation' and industrialisation. A leading group will be set up to orchestrate and enhance the development of the plan. This provides a general frame of intentions, but, more detail on the specifics and the budget will be available in due course.

Such a plan indicates the intention to further strengthen the manufacturing base of China, while making it evolve towards the new techniques, involving internet, ICTs, internet of things and big data/analytics, as well as the so-called '3D printing'. Later in this chapter, we'll come back to China as an internet country.

Specific measures to stimulate innovation in China

There are a multitude of ways to stimulate innovation in China. The most prevalent are tax incentives and administrative regulatory practices.

Fiscal incentives

In China, there is no R&D-related tax credit as there exists in France, for example. On the other hand, there are various reductions on the amount of taxes, as well as preferential tax rates for R&D activities.

One of China's most visible incentives available to innovative firms is a 10% tax rebate (on the normal 25% rate) for 'High and New Technology Enterprises', if they are recognised innovative by the government. In order to acquire such a label, the firm must first prove that it has a patenting activity to a local committee. An additional reduction on corporate tax is granted to firms which manufacture the patented products within China. The resulting cumulative deductions make these measures very attractive.

The certificate is delivered jointly by the ministries of science and technology, of finance and the local tax authorities. It is valid for three years. In effect, this somewhat favours Chinese firms, or those foreign firms which have been in China for a long time and manage their operations much like Chinese companies.

To be eligible, firms must have intellectual property in China, at least 10% of staff engaged in R&D; and 60% of the R&D investments must be carried out in China.

In high technology zones, firms typically enjoy a total exemption of tax for two years, followed by a 50% exemption for three years.

Incubators

It is estimated that there are in total more than 1,300 business incubators in China. Comments on these have been made in Chapter 3 on a critique of the conventional innovation metrics for China.

Incubators constitute a primary place for attracting returnees are members of the Chinese diaspora, who have worked abroad and come back to China. In recent years, the government has found new ways to attract and welcome them. For example, incubators and university science parks have set up offices and extensively equipped laboratories, available to them for reasonable rents. Grants for research, as well as subsidised educational courses and various aspects of counselling, are available for them in such science parks.

The world's largest science park is the Zhongguancun Science Park, covering a large section of north-west Beijing. This zone for entrepreneurship, innovation, education and research, covers more than 230 km². It has an emphasis on ICT, which is in line with the fact that Beijing is one of the world's most powerful places for developing software and computer science. Since its founding in 1988, the park has welcomed more than 20,000 firms, which has obtained a total of close to 60,000 granted patents. Beijing is the primary place for returnees to create a start-up company in China. A majority of them are in the software sector.

This science park includes six sub-parks, which are more or less focused in specific areas. For example, Desheng Park concentrates on incubating small and medium-sized enterprises in the areas of information and communication technologies and industrial design. It encompasses other industrial areas such as Haidian, the 'electronics city' Changping, Fengtai, and Yizhuang. Early in his tenure, President Xi made an attentive visit to the park.

Incubators and science parks grant specific tax reductions to the enterprises, subsidies for research and development activities, on a case by case basis, and also provide subsidised education of a technical as well as managerial nature.

Procurement

Other policies are supportive of innovation in a less direct way. These include procurement, which is one way to promote the development of innovative, domestic ('indigeneous') offerings, and to accelerate their diffusion. In China, public procurement represents 20% of the country's GDP. In the coming years, this ratio is expected to decline somewhat, as the weight of the SOEs is reduced in the economy. Initial policy has been instituted in the 2006 Science & Technology Plan mentioned in the previous section. More specific measures will be enacted in the near future, which hopefully will also reduce discrimination against non-Chinese firms. These are potentially powerful, given the large size of the public sector as a client.

On procurement as an incentive for innovation, the European Commission is following a similar logic, using elements of the North American Small Business Act (SBA) to force government agencies to purchase a fraction of their goods and services from small and medium sized companies. The 'buy Chinese' circular 618, issued in 2009 jointly by MOST, MOF and the NDRC, has continued to attract criticism from the international community, as it contains a long list of items and services purchased by the government, which would preclude non-Chinese firms from many bids.

Standards

Standards are the realm of the Standards Administration of China. Following China's entry into the WTO in 2001, trade barriers such as import quotas, tariffs or licensing limitations had to be dismantled. Standards therefore are a tool for allocating power and innovation within an economy, favouring domestic actors if the government so pleases. In the early 1990s, Europe's example of the GSM standard provides a successful case. It catalysed the emergence of powerful companies, such as Ericsson and Nokia. Years later, these firms are champions no more and have yielded to challengers from Asia.

A classical example in China is the TD-SCDMA standard for telecommunications. This was initially developed by Chinese telecom engineers together with Siemens. It was selected in 1998. Public funds were allocated to its development, and research consortia were created in 2006. At that time, President Hu launched indigenous innovation as 'the core of national competitiveness'. MIIT announced that the standard became the compulsory domestic standard for 3G mobile telecommunications. Technical problems delayed its introduction. As a result, in 2000, 3G licenses for competing WCDMA and CDM 2000 were granted to China Unicom and China Telecom, retaining the largest operator, China Mobile, as the champion of TD-CDMA to give it a chance of survival. By then, however, customers were already moving to 4G, so that, today, the resulting outcome is far from being a resounding success.

As of Spring 2015, China Telecom and China Unicom are making pilot tests of 4G combining TD-LTE with LTE-FDD networks, in 16 Chinese cities. In the coming years, no doubt a multiplicity of services will continue to emerge to serve the now 475 million users of 3G mobile networks.

Regulation

Regulation constitutes another powerful way to favour Chinese innovations. Smart cards, pioneered by Europe for banks transportation, etc, are one such example. In 2007, regulation of the China Compulsory Code (CCC) essentially forced the producers to divulge their operating system. Following a threat of trade actions by the European Union, Japan and the USA, the regulation was modified. Smart cards used in many Chinese applications, however, are still required to comply with CCC certification.

Similarly, in the wind turbine sector, a continuous stream of government regulations and requirements favour the development of the already vibrant sector. The government almost totally controls the market of the 1.5 megawatts wind turbines, which constitute the workhorse of today's wind farms.

In brief, public policies can be designed to facilitate innovation. At the same time, the whole architecture of the 'indigenous innovation' movement includes elements, such as non-tariff barriers, for example, making it somewhat difficult for non-Chinese firms to participate in the Chinese market. This seems to be particularly the case with the recent Strategic Emerging Industries (SEI) initiative.

Policies and public initiatives to reinforce China as internet country

China's first internet connection was in 1987: a connection between Beijing and Karlsruhe University. Since then, the number of Chinese internet users has very rapidly increased reaching well above 800 million users in 2015. In 2008, China was the first country to declare 'internet addiction' a clinical disorder.

The public policies of China have consistently supported the establishment of a strong internet infrastructure, including promotion of the new protocol IPv6, and fostering firms in this sector. China showcased this protocol at the Beijing Olympics in 2008. The NDRC has launched the China Next Generation Internet CNGI as a five-year plan to step up developments in this area. Many academic organisations have joined CERNET, the Chinese Education and Research Network. This links 20 provinces and provides high-speed internet connection.

The State Informatisation Development Strategy, published by the General Office of the Communist Party of China (CPC) Central Committee and General Office of the State Council, gives the long-term ICT development goals up to 2020. Nine key aspects are retained: promoting informatisation of the national economy; popularising e-government; promoting an advanced internet culture; advancing informatisation in education, health care, and public safety; expanding information infrastructures (e.g. wireless broadband and 3G and 4G wireless networks); exploiting information resources more efficiently; improving the global competitiveness of the Chinese ICT industry as a whole.

Another important document for China's development in the near term has designated ICT as a key societal and economic goal for China by 2020. In each of the five-year plans for China's economic development, ICT is a priority. The latest, the Twelfth Five-Year Plan of 2010, designates the ICT industry as one of China's seven key emerging industries.

As of 2015, it is estimated that the broadly defined 'internet economy' already represents close to 20% of China's GDP. An insight into the unfolding of China as an 'internet country' in the coming years was given by a speech of Prime Minister Li Keqiang in Beijing in March 2015. In his speech, he criticised internet speeds in China, underlining that 'he has been in many emerging countries where internet was faster than in China'.

Present ICT policies build on the 2001 adoption of the 'San Wang' three networks (television, broadband internet and telephone) in a single connection. Li Keqiang officially announced an action plan for 'internet plus'. The goal of this programme is to integrate manufacturing with cloud computing, big data, the mobile internet and the Internet of things. The aim is to foster e-services, including e-commerce and internet banking, as well as industrial networks. Furthermore, the intent is to promote a transformation of the economy via 'informatisation'.

This is consistent with the reinforced focus of public policies on information and communications technology. Since 2000 in particular, there has been a flurry of documents stressing the importance of software development, integrated circuit industry, the national broadband plan, cloud computing (many firms provide ample free storage to customers), for which specific government support of a quarter of a billion dollars has been provided in 2011. Smart cities and 'big data' also garner considerable attention and financial support on the part of the central government.

With a strong manufacturing sector, an unfailing support on the part of the government and a very high internet-literacy (in spite of censorship), China seems a natural place for becoming one of the epicentres of the Internet of things. In the coming years, the 'top down' policies and government guidance will interestingly meet a 'bottom up' entrepreneurial and creative movement.

The West is easily obsessed with the nefarious censorship policy coming from the 1993 Golden Shield project, which, in 2000, led to the massive surveillance programme of the 'Great Firewall'. The latter blocked an estimated 18,000 websites in 2014. As a result, the world outside China is not aware enough of the potential of China as an internet country. Following a recent statement from Mr Ma Jiang, Director of the National Statistics Bureau: 'New activities arise from the mobile internet'. It is likely that 'political and societal' censorship may go along with the breakneck speed development of internet services in China.

Regulating fast-growing online financial services

Online financial services constitute a growing and promising segment of internet services. Looking at the near future, the governor of the central bank of China, Zhou Xiaochuan, recently celebrated the development of online financial services. The suggestion was that the bank would adopt a positive attitude to provide for the development of internet finance. This sector still represents a relatively small activity, as compared with 'traditional' banking. It is estimated that it was the largest in the world in 2015, with an estimated number of well over 1,500 lending platforms.

Early in 2015, internet banking in China represented transactions worth about $60 billion, up 22% from the previous year. Mobile banking alone increased by 30% in the same period. It seems likely that this sector will develop extremely quickly in the coming few years. Yu'e Bao ('remnant treasure', connected with Alipay, briefly described earlier) is rapidly expanding. Tencent's WeBank (from WeChat) was launched early 2015, in Shenzhen, by the prime minister, who stated:

> 'We will lower costs for and deliver practical benefits to small clients, while forcing traditional financial institutions to accelerate reforms.'

In this sector the risk of a default is real, so regulators act cautiously. They already have a somewhat relaxed access to funds via the Internet in 2015. The ongoing momentum is powerful.

Regulators have banned the use of the virtual currency bitcoin. For the first time in late December 2014, the China Insurance Regulatory Commission published guidelines for internet insurance, setting the threshold for entry and some safety aspects for selling insurance policies online.

In the near future, it will be interesting to watch what the regulators will do in this area. Traditionally conservative, they may favour the established financial sector, itself in need of substantial reform. This is underlined by the drop of the Shanghai stock exchange, which went down 30% between August 2014 and August 2015, and the manipulations of the authorities in attempts to stabilise it in the course of the summer of 2015. Alternatively, they may use the newcomers to force changes in the banking system, also allowing online banking players to emerge and flourish, while having light but smart regulation to avoid the worst risks. One inconvenience from this would be that it does not create as many jobs as the manufacturing sector.

In brief, when looking at the government and public administration as orchestrators of the policies and practices in innovation, the patterns for promoting China as a global innovator are as follows. First, the pro-science & technology government is relentlessly supporting innovation, in particular technology-intensive, so-called 'indigenous' innovation. Second, the policies and laws (anti trust rules, procurement, utility model patents, regulations, etc. . .) are occasionally used to keep non-Chinese competitors out. Third, there is a strong focus on making China THE Internet country, as a way to create new activities and jobs. In the near future, online banking must be watched carefully.

China possesses an arsenal of programmes, policies and regulations to promote innovation, particularly domestic ('indigenous') innovation. These programmes have substantial budgets and have the full support of the top leadership. One may sometimes wonder how effective is the coordination between them, especially given the complexity of China's administrative apparatus. They also tend to excessively rely on 'mega projects'. In the following chapter, we turn to the patterns followed by firms to propel China into the status of global innovator.

Chapter 9

Becoming a global innovator – patterns in firms

In this chapter, we first look at companies in various industrial sectors, illustrating China's business environment of speed, fierce competition, and demanding customers. Depending upon firms and sectors, this also highlights various approaches to innovation, from acquiring technology and knowledge, to adapting business models and creating their own capabilities internally. We then look at the patterns that inform the ways with which Chinese companies are taking a journey towards making China a global innovator.

From acquiring to creating: China as a world leader in the ICT industry

China is already leading in many areas of the mobile telephony, gaming and internet-based offerings. This is the case for services and software as well as for hardware devices. On the other hand, it lags behind in the crucial area of microchip design and fabrication. In this area, for the foreseeable future, the country is likely to continue to be heavily dependent upon foreign suppliers, from Japan and the USA in particular.

Beijing is a world centre in ICT and mobile internet, a sector which counts 6,000 firms in China. Chengdu, the capital of Sichuan Province, leads in the gaming industry. More than 1,500 ICT and software companies produce more than $5 billion in revenues.

One promising newcomer is the game company Tap4fun. Initially called NibiruTech when it was founded by 12 people in 2011, it now has 350 employees and is making games such as 'Spartan Wars.' Many of its millions of players are not Chinese. Its studios are in Chengdu and Paris.[1]

As is often the case in the mobile internet innovation, the business idea is often not ground-breaking, but provides a novel way to extract value from an activity when diligently implemented. This entrepreneurial skill is extremely well developed in the Chinese population. Two companies exemplify this dynamism: Xiaomi and Tencent. Both are in the area of internet/mobile telecommunications. More than anywhere else change is fast and often brutal.

A fast rising newcomer: Xiaomi

Xiaomi means 'little rice' in Chinese. This company sold its first smartphone in 2011, 20 million in 2013, and 60 million in 2014. Headquartered in Beijing, it was founded in 2010. Its devices are only sold online. The handsets are produced by the Taiwan-based manufacturing firm Foxconn. Xiaomi sells services as well as devices. The phones have a customisable interface which is updated each week, responding to users' input via the Internet. The company's founder, Lei Jun, has a passion for mobile phones and claims that his company's success is not that it sells phones, but 'an opportunity to participate'. As of summer 2015, the company has 70 million co-creating users. Relying on its customers, Xiaomi does very little advertising. It recently redesigned smartphones into 'smart TVs', sold for RMB 3,000 (€365) starting mid-2014.[2]

Fast moving Tencent

The giant company Tencent had revenues of $9 billion in 2013. It has 300 million users for messaging, shopping, banking, buying services, but overall, Tencent's services are used by more than 800 million people. These include the instant messaging system QQ, which is a clone of ICQ, an Israeli invention, acquired by the North American firm AOL. Tencent charges only small sums for its services and is also known for having created its own online currency.

WeChat is a good example of Chinese innovation. Launched in 2011, it is a free social networking platform akin to WhatsApp. Developed by Tencent, it had 500 million users in 2014.

Alibaba, Baidu and Tencent are currently China's 'big three' online operators. Tencent is fairly oriented towards the USA. It now designs offerings for 3G available in more than 300 Chinese cities.

China's e-commerce is perhaps the world's most effective and efficient. This is partly because the Chinese retail system is not working very well. It is remarkably inexpensive and efficient to have same-day delivery in most cities. Finally, the young generation has a lust for the Internet, as discussed below in this chapter.

Today's models of entrepreneurial innovation for the future are the CEOs of the above companies. Alibaba's Jack Ma is particularly well known after its introduction to the New York Stock Exchange in 2014; Pony Ma is the CEO of Tencent, and Lei Jun leads Xiaomi.

Electronics hardware and the 'makers' movement'

The activities based on information and computer technologies offer great scope for growth and efficiencies, just like elsewhere in the world, but China is forging ahead in other areas as well.

The Shenzhen region with its numerous companies supplying high quality components, goods and services, is probably the best area in the world to manufacture devices of all kinds. Most of the world's digital devices are manufactured in the district of Futian. Small and nimble companies are able to produce prototypes or small series of devices very quickly. One example for this buoyant, high-quality performance is Haxlr8r, an 'incubator/accelerator' for new ventures founded in 2011.[3] This region is densely populated with entrepreneurs leading firms for quality manufacturing, including the giant 200,000 employees manufacturing city Foxconn, which invests 1% of its sales in R&D, compared to 11% in the case of Apple. However, it files more patent applications each year than Apple. The Shenzhen area is also home of the contract manufacturer Seed Studio. Created in 2008, it is one of the world's largest manufacturers of open-source hardware.

Huawei – the poster child of Chinese innovation

Huawei has become the paradigmatic Chinese technology company after it shook up the global telecommunications industry. It was the first truly high-tech company from China that globalised successfully. Huawei has more than 140,000 employees, revenues of $46 billion, and R&D investments of $5 billion.

Apart from being a feared telecom innovator, it also stands for organisational innovation. Founded by Ren Zhengfei, a retired Army general, whose connections helped the company to dominate the Chinese market before going overseas, there was always the lingering question of how this company will transition into the next generation of leadership. Huawei invited several foreign consulting firms and introduced tried-and-tested processes and organisational designs. At one point, when it became clear that it needed more diverse blood, it specifically hired foreign engineers. Later, it targeted foreign executives to take over key transformation positions inside the firm. But as this was still not enough, it instituted a triumvirate of top leadership: three senior executives (Guo Ping, Ken Hu and Eric Xu) each take turns and rotate through the chief executive position, each assuming the acting CEO role for six months, while one of the others serves as president of the firm, and the third retains the position of chairman on the board of directors.

Thus, all three executives are always closely involved in both strategic and executive decision making, and are ready to take over once it is their turn at the helm. This triumvirate leadership is said to be more suited to the dynamic and fast-changing environment in which Huawei competes. However, the West can take at least credit for the original idea, as it was inspired by a book on business leadership called *Flight of the Buffalo*, written by James Belasco and Ralph Stayer.

In the course of 2015, Huawei has developed its presence in the mobile phone market, producing more then 2,000 devices per day, with a 99.5%

success meeting the quality targets in its Shenzen plant. The company plans to sell more than one million units in 2016. This illustrates that China companies, such as Huawei, can deliver on bold gambles.

The 'makers'

The region of Shenzhen is a natural environment for the 'makers' movement. Makers are 'tinkerers' who design and build new devices from existing components or accessories. You may thus end up with a Swiss army knife, which is also a portable TV set, or a flashlight, which also functions as a corkscrew. The movement has also created a cell phone specifically for Chinese migrant workers. It makes use of 3D manufacturing, sometimes also called 'additive printing'. One of these 'hackerspaces' is called XinCheJian, meaning 'new factory'. The firm Win Sun has used a 6 × 10 metre 3D printing machine to produce elements of a three-storey building in two days. The elements are made of a mix of fibres, steel, cement and other additives.

The commercial impact of this movement has been low so far, but one could imagine some products taking off and being sold online. There have been several 'makers fairs' in China. The first one took place in April 2014 and was attended by tens of thousands of people. The main point is that the 'makers' represent a strongly *bottom up* dynamic.

As the 'makers' movement incorporates open source design and dimensions of fashion, it has been adopted by educational programmes in leading universities, such as Tsinghua or Tongji, as part of their educational programmes. Such universities provide laboratories (not just the infrastructure and equipment, but also the entrepreneurial embeddedness) which act as playgrounds for the creativity of diverse students to work together and blossom. Tongji University in Shanghai is known for its extensive building and housing of design and artistic creativity.

A bottom-up movement

In early 2015, Prime Minister Li Keqiang paid a visit to one of the 'maker's spaces', thus bringing instant attention and encouragement to this 'bottom-up' approach. The acts of the top leadership are always highly symbolic and carefully observed and interpreted. A visit such as this one, in a country as hierarchical and hungry for the next big thing, meant that the 'makers' movement became an instant hit. This has practical implications: in this *élan*, it became suddenly much easier for makers to find space to open new laboratories. This illustrates the pattern of a strong bottom-up movement, acting as a sociological counter-action to a highly vertical, omnipresent, but ready-to-experiment government. Contrary to old-fashioned Western views, the Chinese government does not need to force its will top-down through the bureaucracy, it just unlocks the gates, often with little effort but to great effect, to allow the streams of energy and creativity to flow from the bottom up in new directions.

This also illustrates the astuteness of the regime. It courts and includes this young, creative generation and welcomes this sociological trend into the Communist Party. This sends the message that it encourages creativity and innovation and sees them as powerful engines for the country. All this takes place within a very top-down, hierarchical system.

The pharmaceutical sector

China represents one of the top three healthcare markets in the world, growing at more than 15% per year. The overall healthcare market is expected to reach $1 trillion in 2020 (i.e. three times what it was in 2011). This robust growth is expected to continue because of a number of factors: rapid growth of the already large middle class, strong urbanisation orchestrated by the government, an aging population, and growing public support for healthcare.

China's drug market is highly fragmented: the top ten pharmaceutical companies have only 10% of the total market, whereas this ratio is 50% in most Western countries. In 2013, the pharmaceutical industry represented $100 billion in sales. In the last five years, the government invested $50 billion per year in this sector. Several regions of China compete for leadership. In 2012, more than 60 partnership deals were confirmed, mostly with non-Chinese firms.

By and large, however, in life sciences China is lagging behind OECD countries; it is indeed hard to say by how much, but depending upon the area, one can roughly estimate 10 to 15 years.

China has a dual medical system. One of them is the well-established Traditional Chinese medicine (TCM), grounded in centuries of practice-based experience. In the 1950s, China made a great effort to infuse more system and structure into TCM so as to retain an independent Chinese way of medicine and allowing it to function alongside the science-based Western approach, the other dominant system in China. It heavily relies on drugs, for which China is already the world's third largest market, with growth rates close to 20% per year.

Back in 2003, Sibiono, a company founded in Shenzhen in 1998, made history by obtaining approval for a gene therapy drug against head and neck cancer, Gendicine. Sibiono is no longer alone. Many similar companies are emerging, made possible by highly experienced 'returnees', who were attracted by the opportunities and talent available in the country to come back and lead companies in this sector. One estimate is that over the last five years 80,000 life-sciences professionals have come back to China.

Shanghai-based Lide Biotech is active in the oncological area. The company provides validation and testing of gene therapy approaches for the treatment of specific cancers. It was founded in 2011 by Dr Danyi Wen, who spent several years in the USA and came back to China in 2007. Other firms founded by senior returnees include Hua Medicine, Ascletis, who started with a record funding of $100 million, and Beigene in Beijing, which concentrates on diseases such as liver and gastric cancers: frequent in Asia, they are somewhat overlooked by the West.

Another example is BGI, formerly called the Beijing Genomics Institute. In 1999, a team of scientists left an affiliate of the Chinese Academy of Sciences to found what is now the largest biotech company in China; it does DNA sequencing on samples received from all over the world. Its innovative approach is to boldly rely on the typically Chinese approach of mass manufacturing, using close to 200 sequencing machines. It currently has 4,000 staff and generates more than a quarter of the world's genomics data, more than any other institution in the world.

Medical devices

In 2016, China's medical devices sector is expected to grow to $44 billion in sales, and all indications are that it won't stop there. The use and application of medical instruments is driven by the healthcare sector, and is related to overall growth in e.g. traffic (more accidents means more reconstructive surgery), aging, dispensable incomes, etc.

Mindray

Founded in 1991 and headquartered in Shenzen, Mindray is an example of China's push towards global reach. This company produces diagnostic and monitoring systems for patients, as well as X-rays, ultrasound and MRI machines. It also has a veterinary division. It is present in 22 countries with 7,800 staff, had a sales revenues of $1.214 billion in 2013 and has been listed on the New York stock exchange since 2006. China represents 46% of its business, but its fastest growth (20% in 2014) is in Europe.

In the presentation of the firm, the word 'innovation' appears in the first paragraph.[4] Indeed, it invests close to 10% of its sales in R&D, employing more than 1,800 engineers. Typical of so many 'technical' companies in China, Mindray is now making and selling products as good (or even slightly better) than those offered by its Western competitors, only for a fraction of the cost. It is now poised to forge ahead and to come up with pioneering devices, features and services. The question is not whether it will happen but when; we expect this to happen in the very near future. With the recent acquisitions of Wuhan Dragonbio (orthopaedics) and Datascope (patient monitoring), Mindray will rapidly have a world-wide reach, in a market which is so patently global.

While China wants to be present in the so-called 'promising' key industries of ICT, health, advanced energy and materials, it also wants to have a strong impact in more 'conventional' areas, such as automotive and aircraft, as discussed below.

Car industry and self-driven cars

China is already the world's largest market for automobiles and the largest car manufacturer. No wonder that automotive investors consider the country as a priority. Globally, the industry is more than 100 years old and characterised by mergers and acquisitions. In China, the industry is marked by state-owned enterprises and their joint ventures with foreign firms such as GM and Volkswagen. But some local firms have different origins. Most people recognise BYD, the battery-maker turned into a company producing electric cars and buses. New start ups start small, so it is always eye-catching when someone tries to enter at the system-integrator level, such as Qoros.

Qoros

With investments of $1.5 billion from Israel Corp. and the local Chinese auto maker Chery, Qoros was founded in 2007 with mostly European executive management and Chinese staff. Two of its design centres are in Munich, Germany, and Graz, Austria. Its first sedan car was launched in late 2013 in China, followed by a progressive introduction in Europe.

The Chinese name for Qoros combines two words meaning 'learning' and 'excellence'. Two 'innovative' characteristics are attached to this firm. First, on the dashboard, a panel provides a highly digital control of the car functions; the panel has been specially designed for Qoros by the German firm Hartmann. Second, the design takes into account the global specifications on emission, safety, etc., as well as a common platform, which can accommodate many different car models. In this way, it is planned to introduce a new model every six to twelve months. The ambition is to build a brand with a strong German flavour, designed and developed for the world's most dynamic market, with the goal to produce the first truly global cars for metropolitan families. It has a plant in Changsu which can produce 150,000 cars per year, and even though sales are still lagging behind the founders' high expectations, the firm hopes to be profitable by 2018.

As the Western world looks somewhat patronisingly at China's automotive industry and their ambition to carve out a pioneering world role in the industry, it may be useful to remember the case of Korea. In the 1980s, Korean cars were technologically backward, not attractive to the world markets, and based out of a fairly small and protected country. Thirty years later, European car makers are developing some of their new models in their technology centres in Korea, flatly contradicting the conventional wisdom that '. . . manufacturing may go to Asia, but Europe will retain the high-value functions of design and R&D/innovation'. With its talent, financial muscle and powerful market, China should be able to climb this learning curve just as Korea did, possibly within ten years.

One area in which China is expected to have a strong innovative impact is self-driven cars. For this purpose, BMW has teamed up with Baidu in China. Another relatively small car maker (roughly 0.5 million cars produced per year), Geely/Volvo has made bold announcements in this area, too.

This constitutes another pattern of innovation in China. Industrial sectors considered as more 'mature', such as the car industry, are suddenly again moving fast and adopting new approaches, often involving ICTs. This phenomenon draws on a massive manufacturing base, a large and demanding market, hungry for technical novel features, but facing specific acute situations, such as severe traffic congestion in cities, that serve as unique impetus for innovation.

For the near term, electric cars are perceived as a solution to pollution in cities, when in fact it merely displaces the source of pollution: electricity still has to be produced somewhere; not all of it is produced using renewable sources. Truly green solutions would seem to include novel approaches to public transport, or walking and cycling.

There is, however, a big push to produce electric cars in China and no doubt this will be an area for multiple innovations. While existing manufacturers are struggling, such as California-based Tesla in the more expensive ($100,000 per car) segment, new actors are jumping in. The electronics manufacturer Foxconn announced in 2015 that it is investing $800 million to produce this kind of vehicle. At the same time, it announced a partnership with Tencent, possibly for self-driven cars.

Such corporate shifts are made possible by the large and dynamic economy of China. Qoros itself is, as a firm, an innovation, since it could be labelled with the unusual term of 'corporate start up' (i.e. a firm born from scratch, as a global corporation, with a massive initial investment). Another such 'corporate start up' is COMAC, in the aircraft industry, on which we focus are attention next.

An emerging civilian aircraft industry

Systems integration is the key word for the design and building of airplanes. This is an area in which China is badly prepared, as a result of its traditional way of doing things: excessive respect for hierarchy, lack of communication and sharing, etc. Yet, for strategic reasons (national independence), for national pride, and also because of the forecasted huge market (more than 5,000 planes with more than 50 seats needed in the country by 2030), China decided to create a civilian aircraft industry. The Tianjin free trade zone is home to various joint ventures between Chinese firms and companies such as Airbus or Boeing. Airbus considerably expanded its presence in China as early as 1985.

In China, the Aviation Industry Corporation of China (AVIC) is dedicated to the development and manufacturing of aircraft engines. As a first step, it focused on piston engines for civilian planes. AVIC obtained technical know how from the German-based company Thielert. It also has a joint venture with MTU, the only independent aircraft engine manufacturer in Germany not connected with the trio of engine suppliers, Rolls Royce, GE and Pratt & Whitney. AVIC also acquired American aircraft engine manufacturer Continental Motors in 2010 and the US aircraft maker Cirrus in 2011.

Along with other large Chinese SOEs, AVIC co-founded COMAC in 2008 to focus on designing and building large commercial aircraft. In 2010, COMAC's capital was close to $3 billion. Within two years it had built a plant and offices for 50,000 employees south of Shanghai Pudong airport.

The successor to the ARJ21 aircraft, the larger C919 jetliner is a 170-passenger plane, competing with the Airbus 320 and the Boeing 737. It is anticipated that the C919 will be delivered to airlines no later than 2020. Wide-body airplanes are proposed and under design. By committing to the new CFM 56 LEAP engines of the long standing joint venture between the French company Safran (formerly SNECMA) and the US General Electric, COMAC had a defining role in triggering the go-ahead for the development of this new engine.

Despite the huge technical challenges for COMAC in developing new aircraft, the innovation is not in the airplane but in the redesign of an industry. COMAC must somehow overcome the reluctance of Boeing and Airbus to accept a new competitor. After all, products' life spans and ROIs are measured in decades, and the aircraft industry needs stability to be profitable. In 2011, COMAC signed a long-term cooperation agreement with Canada's Bombardier, and will have access to the avionics elements necessary to build the airplanes.

The other area of innovation is in the recruitment and training of staff. The company has engaged in a very active programme of hiring non-Chinese engineers, who then rapidly train the Chinese. An element of that is China's 'Thousand Talents Programme' in which COMAC is also heavily involved. A key innovation at COMAC is the setting up of a system of workshops, tutoring and training courses, to boost the competencies required in this industry.

So even in this sector, which looked so settled and unassailable, China has innovated in another, fundamental way. For the first time in the history of the world's aircraft industry, a technology transfer programme allows a most critical component, the wings of an airplane (in this case the A320) to be produced, in China, outside the country of the firm building the plane.

High speed trains and technology transfer

Drawing on some $300 billion of investments, China today has a world-class capability in high speed trains, in spite of the tragic and calamitous 'Wenzhou crash' that killed 40 people in 2011. This was apparently the result of a faulty signalling system, which included on-the-ground as well as on-board elements.

Leveraging astutely combined technology transfers from the French TGV, German ICE and Japanese Shinkansen, China has built a state-of-the-art industry, without having to wage any patent litigation in the process. Currently, China represents half of the world's investments in high speed trains. By 2020, China plans to have a network of 25,000 km of high-speed tracks. A specific, practical innovation in this field is the fact that the Harbin-Dalian high-speed railway can operate in extremely low temperatures, a first ever worldwide.

Similar to other sectors of activity, the goal is to reach international markets. In July 2014, the first Chinese-built high-speed train arrived in Istanbul; however, projects with Thailand and Mexico have been cancelled.

Back in 2005, Siemens was invited by the China National Railway (CNR) to join a bid for the 115 km Beijing to Tianjin high-speed railway. The first trains were built by the German plant; the following 57 trains were built by the CNR plant in Tangshan. Siemens trained more than 1,000 CNR engineers in Germany. The line opened in 2008 before the Beijing Olympic Games.

Other industrial sectors

It is beyond this book to exhaustively cover here all industrial sectors in China. This would require a study of real estate, where so many large fortunes have been made, retail, and food, where there are huge companies, as would be expected in such a populous country. But many of these sectors do not seem to be likely to generate a large innovative corporation capable of becoming a global player in the near term. One company that does seem closer to that is Sinochem.

Sinochem

Founded in 1950, the company had 50,000 employees and revenues of RMB 460 billion in 2013. With a vision to be a global actor, one objective is to rely more and more on 'technology leadership'. Indeed, its history provides a classical model of climbing the value chain, which in turn implies the need for more innovation. Coming from import-and-export, the firm's focus moved to the petrochemical trade, and once it had become a multinational company with a global presence, it expanded to energy and real estate. The current phase demands substantial investment in R&D, working with many external partners and cooperating with companies such as Monsanto in the USA and DSM in the Netherlands. This path much resembles the trajectory followed by the oil and petrochemicals company Sabic, from the Kingdom of Saudi Arabia.

Not surprisingly in a country which has been seized by building fever, companies in the area of construction equipment have had an opportunity to emerge and become strong. Two such companies, with boldly innovative products are highlighted below.

Sany: equipment for the building industry

Sany specialises in pumps for concrete, a useful specialty in a country that builds so much. It also sells trucks and excavators. Founded in 1989 in Changsha, the capital of Hunan province, it now has 90,000 employees. Its sales total more than $50 billion per year. It currently invests 5–7% of sales in R&D, with centres in Germany, India, China, the USA and Brazil. Sany produces rather low-end machines and has a local distribution system, selling to leasing companies, whereas the conventional firms sell to construction companies. Sany's fame comes from its development of a pump which can push concrete up to a record 72 metres up. This bold challenge was made possible, in large part, by the integration of considerable technical know-how gained in the 2012 acquisition of the German Mittelstand company Putzmeister.

A bold crane manufacturer

Another example of bold innovation is the crane manufacturer Zhenhua Port Machinery Company (ZPMC). Mr Guan, a bureaucrat who in 1992 retired from the government, set up a private company manufacturing cranes for the rapidly expanding port of Shanghai. Before actually producing even a single crane, Guan convinced the Port of Vancouver to do an advance purchase. Quality, here also, was a company value and cranes were sold to many Chinese ports. Even for its international business, in order to retain the cost advantage, Guan decided to build the cranes in China and transport them to their destination on ships. This had never been done before.

Success was achieved when the firm cracked the market of the port of Hamburg. By 2010, the company – now called Zhenhua Heavy Industries Group – had more than 60% of the market world-wide for large harbour cranes.

Services

Between 1978 and 2010, the share of the services sector in China's GDP has more than doubled to reach 40%, and now stands close to 45% in 2015. Extrapolation is fraught with difficulties, but it stands to reason that this share will continue to grow, offering enormous opportunities for innovating in this section of the economy, above all in on-line supported services.

For instance, online financial services may undergo a boom. The government will, very gradually, deregulate the sector. It has already allowed consumers to open several online accounts instead of just one. Interest rates are regulated, but as you can imagine, there are ways to circumvent these limitations. The government's fear is that once there is another financial crisis (it's a bit like the public bus system: you know there will always be another one, you're just never sure when)[5] and consumers lose money, poorly managed public interest rates will significantly add to public discontent. In any case, China is extremely cautious, and perhaps rightly so, about the Wall Street system. Following Alibaba's Yueboa, discussed elsewhere in this book, other players are coming in and offering services at very low cost. Again, cost innovation is the operating approach. Furthermore, numerous start ups are developing new businesses in various financial services. This includes leveraging 'big data/analytics'. This sector is due for a rapid expansion, globally as well as in China. The innovation will not be in the financial products themselves, but in providing customers with services at minimum cost by leveraging over a large number of customers.

A totally different area is the food services industry; one example below is in the restaurant business.

Food and entertainment: Hai Di Lao Hot-Pot restaurants

Founded in 1994 in the province of Sichuan, this rapidly growing chain of restaurants now has 15,000 employees working in 75 restaurants in 16 cities.

Its cuisine draws on the rich diversity of Chinese cooking. Its business model involves an industrialisation of the reparation of the ingredients. It also is characterised by an entertainment ingredient; either TV programmes or karaoke, as well as some degree of preparation of the dishes by the clients themselves. The entertainment is not exclusively focused on the dining (other companies do that already to a great extent, so no differentiation), but includes the inevitable waiting period between arrival and being led to a table.

Besides pre-ordering, Haidilao offers game rooms/tables, internet stations, children's entertainers, etc. In another innovative move, it has conference dining rooms which can be booked in two Haidilao restaurants. While we don't think this is how most people would want to eat dinner, it is certainly a novel way to combine the necessary with the social (which is also a necessity for successful business in China) when two business parties are in remote cities.

Innovation in China's global firms

China has numerous large companies. Many, in the food sector, for example, are very large but concentrated on the national market. A small number of them have global reach and fame: Alibaba, Haier, Huawei, Lenovo and ZTE.

Chinese executives often underline the uniqueness of the management of China's companies, as compared to non-Chinese ones. If challenged on this statement, the argument usually resorts to the combination of the soul of Asian culture with Western management practices. The beautiful thing about leadership and management is that so much can be said, one thing and its opposite, while so little is truly new … Common sense elements, such as talent, motivation and good judgement remain at the crux of the matter. Unique to China is a generally relentlessly fast and efficient execution in managing firms.

As in the West, innovation is a staple term in China-based global companies. For example, Haier's Senior Director Global R&D Taoming Wang states that '… a fundamental ingredient for our future success worldwide is innovation'. How can we put this into practice in the so-called 'conventional' industrial sector of white goods and home appliances?

Haier

Haier, a manufacturer of white goods and home appliances, has 80,000 employees and sales of $30 billion. In 2014, it had the highest world's market share in white goods: 10%. Such result means that the products present a good price to quality ratio. It does not mean that current Haier products are innovative.

In 1920 the Qingdao-based company started as a producer of refrigerators. Like so many Chinese companies, it became state-owned after 1949. In 1984, the city called a young manager to become its director, Zhang Ruimin, who soon was to become part of China's management folklore by asking workers to smash its faulty refrigerators. Quality became central to his crusade, making it the centrepiece of his marketing philosophy.

It was quite a challenge to build this mantra for quality among plant staff who had gone through the utter disorganisation of the Cultural Revolution. Zhang drew on Japanese practices of relentless Kaizen, (i.e. constant improvement). Daily performances were evaluated and salaries and awards were allocated accordingly at the end of each month. This expensive process was deemed necessary to create the desired drive for quality and constant improvement. The objective was to build a brand based on quality.

This was followed by the creation of autonomous units, the so-called ZZJYTs, to ensure a close connection between the customers and the firm. The slogan was to have 'zero distance to the customer'. The Western concept of 'empowerment' could be used to designate this movement.

In 1992, the firm avoided bankruptcy by raising capital on the Shanghai Stock exchange. It made an initial public offering (IPO) for 43% of its refrigerator business.

As an indication of its global reach, Haier has made substantial investments in an international array of R&D centres in China, Japan, Australia, Europe and the USA. Haier's investments in R&D/Innovation, however, have not yet resulted in truly distinctive innovative technology. It is in close competition with Arcelik, a company from Turkey, another 'emerging' economy. One 'innovative' feature of Haier is the care taken by that company to deal with after-sales service.

One recent success of the firm was the development and sale of a wine cooler. Haier was able to design a quality product in such a way that production costs were a fraction of its competitors. Haier now enjoys a 60% market share in the USA for this item.

The amount of investment by China outside the country has steadily increased. In 2013, this flow was similar to that being invested by foreign firms in China. We stand at the beginning of a large wave of mergers and acquisitions by cash-rich Chinese firms, with the aim of buying more technology and developing global brands. Hopefully, both the investors and the receiving countries will behave responsibly and exercise good judgement. Often, going global comes with occasional bumps in the road, as in the case of Suntech below.

Failure to globalise: SUNTECH

Founded in 2001, Suntech was at one time the world's largest producer of photovoltaic solar cells. It has its headquarters in Wuxi and was listed in the New York Stock Exchange in 2005. At the time, it was celebrated as a top innovator in its industry. Its R&D counts 350 staff on three continents. However, state support for the market waned and prices plummeted. Having expanded too fast, the company filed for bankruptcy in 2013. It was purchased by a smaller rival, Shunfeng Photovoltaic.

Currently, China's plan is to increase the production of solar electricity from 28 gigawatts in 2015 to 78 gigawatts at the end of 2017.

When it comes to global innovation, China puzzles established international rankings organisations such as IMD's Competitiveness Report, CASTED, the WIPO Global Innovation Index and the EU index. Depending on the exercise, China ranks between positions 64 and 21. Without giving them more importance than they deserve, these rankings generally point to the fact that, according to certain criteria, China is progressing but still has some way to go. Note, however, that when Japan was enjoying its decade of global economic leadership in the 1980s, Western rankings did not rate its innovativeness very high at all.

Ranking systems also exist for firms. According to a ranking by Booz in 2008, of the top 1,000 companies with the most R&D investment, 15 were Chinese. Four years later, there were 47. In the same period, the number of Indian companies only increased from four to nine.

Such rankings do not constitute an exact science. They do not predict a country's success. Certainly, they do not provide absolute answers, or provide much insight. They are useful in that they encourage readers to challenge them by raising questions. They offer a tool for reflecting on what makes a country innovative, by evaluating and discussing the choice of their criteria, which, by the way, are Western. Such discussion may be useful in refining the patterns, followed by China on its way to becoming a global innovator, as discussed in the next section.

Patterns in China becoming a global innovator

Investing massively in R&D is not the key reason why China is becoming an innovation powerhouse. Most of the highly successful firms mentioned earlier pale into insignificance compared to other, globally more established firms, in terms of accumulated R&D. Very rapidly adapting/innovating to reduce cost, in China's vibrant and growing markets, is all that has been needed so far, and this is still the current pattern in many sectors.

In the recent past, large technology-intensive firms have begun to invest much more in R&D/innovation. In 2012 they accounted for a large fraction of the R&D investments of the country. The large manufacturing firms generally have plenty of cash to do so. They do not invest yet much of their effort in R&D that might lead to long-term technical breakthroughs, instead investing mostly in developing technology. The shift will soon take place in that direction, and much learning has to take place to make this transition successful. For the time being, firms focus on downstream developments, which they execute with remarkable speed and by involving extensive external collaboration sharply focussed on the market. This entrepreneurial approach has been called 'distributed innovation' in the past.

Chinese companies have understood that research is cheap, and far from the market, while technological development is quite expensive. Thus, they focus on cutting costs on the latter, while being extremely focused on the market.

Looking at the Chinese scene in the previous chapters, a number of factors are in favour of or against 'Innovative China'. These are given in Table 9.1. Navigating these factors in the near term, China is expected to follow some proven patterns that we outline in the following sections.

The key patterns along which China is moving towards becoming a major source of innovations for the world are: relentless market-orientation and entrepreneurial spirit, fierce competition among domestic firms, excellent orchestration and fast execution of myriads of small-step innovations, drawing on a massive manufacturing sector and efficient supply chains.

Innovating to reduce cost: relentless 'copy and improve'

As has been illustrated in the majority of the examples used, Chinese firms show an almost uncanny ability to copy a product, while both

Table 9.1 A Balance Sheet of Innovation Capital in China.

In Favour of Innovation	Against Innovation
• High entrepreneurial spirit	• Bureaucracy
• Very strong market-orientation	• Hierarchical society
• Talent to extract value from any activity	• Endemic corruption
• Very strong manufacturing base	• Low teamwork and capacity for
• Very motivated by commercial success	systems integration
• Agile and very fast execution	• Censorship and information control
• High internet sophistication	• Acute short termism
• Eager to experiment	• Low investment in long term research
• Robust debate and open society among young urban Chinese	
• Chinese firms fiercely compete and copy each other	

improving and adapting it (i.e. taking out useless features and adding those to suit the Chinese market). This is how China developed Weibo, modelled after Twitter.

This approach to innovation and adaptation to reduce costs is remarkably effective in China. It involves a large number of often minute changes, including replacement by cheaper materials and components. One example is Xiaomi, which grew extremely fast by offering devices and services presenting outstanding price-to-quality ratios.

The 'West' sheds crocodile tears at this so-called 'counterfeiting culture', overlooking the fact that copying and reverse engineering are common practice among non-Chinese firms, too. Western pundits went through the same lamentations about Japan and Korea in the 1970–1990 period. Going back in history, in the nineteenth century the French and the British were bitterly complaining that the then-emerging USA was stealing their technical know how. As the French say, '*plus ça change, plus c'est la même chose*'.

The practice of 'copy and improve' is not primarily targeted at the West. It is aimed at any offering that sells well. Tracking customer preferences with small improvements in simple design and format made to the previous generation of products are critical to commercial success but often do not justify a patent. Thus, as discussed in Chapter 3, output of patents does not reflect properly the ability of Chinese firms to extract value by astute changes in their offerings.

Importantly, within China, firms copy each other relentlessly as a normal way of life. This means that once a firm begins innovating in a given industrial sector, its competitors diligently copy the pioneer. This is an important reason to believe that once a firm starts innovating, the momentum of innovation in this sector steeply increases and maintains itself. This is already the case in the areas of internet, mobile communications and electronic games.

A country fully engaged with the Internet

Walking through the streets and the subway stations of Shanghai, one is likely to bump into youths walking like zombies, watching a film, or a TV 'sitcom' on their hand-held TV. In the subway, passengers are either speaking to their handsets, sending/reading messages, or watching/ listening to a TV programme. In China, well over three-quarters of people connected to the Internet do so via hand-held devices.

This extensive involvement may sometimes be annoying, but it means that in cities the population is fully literate about mobile telephony, so-called 'social media', games and hand-held devices. Much of that rests on home-built companies. For instance, the search engine Baidu was used by close to 66% of China's internet users in 2013.

With revenues of more than $215 billion in 2013, China is the world's second largest – and the most dynamic – market for online sales. In addition, 75% of young Chinese inform their peers about their purchase (clothing, electronic gadgets, etc...) within two days using social media. This provides firms with a novel way of marketing their offerings, while doing real-time monitoring of customer satisfaction. The producers and distributors of so-called 'fast moving consumer goods' (FMCG) thus have a novel and potentially powerful tool for their 'brand-building' and market research.

This also means that a successful product may well rapidly grow in sales, but will be quickly supplanted by another. A pattern for the coming future is therefore that the lifetime of commercially successful products in China is anticipated to be shorter than anywhere else.

At the same time, the Internet is the object of censorship by the government. As we have seen earlier, the 'Great Firewall of China' prevents access to websites considered undesirable by the Chinese government. This censorship concerns opinions or facts, and is indeed blocking a free flow of information on societal issues, but does not seem to be in the way of doing business or technology transfer.

As young, Internet-savvy people become managers in firms, they will fully incorporate the ICT dimension into any business model in which they may be involved. This constitutes a real asset for ICT-intensive innovation, as we only begin to learn how to best use ICTs as enabling elements of business models.

We are still at the beginning of the so-called 'digital revolution'. Its impact will continue to grow in the coming years, going well beyond the business world and affecting all areas of activity in major ways.

China seems remarkably well placed to seize the benefits of such radical and epoch-making changes and has already used the Internet to develop innovative business models.

Relentless market orientation and entrepreneurial spirit

On track towards 'Innovative China', the crucial innovation engine is built around China's relentlessly market-orientated, entrepreneurial drive, with a pragmatic knack for effectively extracting value out of any activity, as well as a willingness to experiment combined with internet-savvy urban customers who are extremely demanding about receiving good value for money. All of the firms' processes are implemented with breakneck speed. The firms' actions towards innovation are strongly supported by public policies, while being practiced with a diversity of approaches by various provinces.

Puffs of commercial success

Crucially, and more so than in the West, China's young entrepreneurs and managers view innovation as very pragmatic, utterly market-orientated and profit-driven. In their view, the importance of the technical element in innovations is much less salient than in the West or in Japan or Korea, for that matter. As has been seen in various examples, innovation involves a host of minute adaptations and modifications/improvements, fast experimentation, relentless execution, drawing on the low cost of doing business and a highly efficient supply chain. On the latter point, both Chinese and Japanese firms tend to put their engineering departments close to the plants, often in the same building.

In the 1980s, the West discovered with great interest Japan's practice of consumer electronics firms conducting early testing in the market and experimenting with products in Tokyo's electronics city Akihabara, in order to evaluate customers' responses and rapidly learn from it. The West did not do much with this discovery, but managers of Chinese firms have the same taste for breakneck speed and experimentation. All this occurs in a large, vibrant, highly competitive market. It is difficult to imagine an environment more conducive to small-step innovations.

In the short term, given the fact that innovations are mostly composed of myriads of small changes, and that Chinese companies fiercely compete with each other while consumer 'fads' spread like wildfire, it is likely that the commercial success of new offerings will be essentially short lived, the latest offering being rapidly replaced by a competing one from another firm. One can thus envisage that new offerings will be introduced at even greater pace, but that they will only enjoy a relatively short-lived puff of commercial success. This is expected to be particularly true for mobile internet: hand-held devices, games, online services, and e-businesses that spill over into the real-world supply chain.

From centres of excellence, spreading innovations to other sectors and regions

Chinese firms tend to fairly 'naturally' integrate product innovations with those in the service area, as well as new ways of doing business (the so-called 'business model innovation'). Several of the above examples illustrated this, Tencent in particular. There is no panacea or simple slogan to define China's innovation practices. In many ways, these still have a long way to go, particularly in the life-sciences sector, but, on the other hand, China moves fast... The firm Xiaomi is an example of these characteristics in the present period and is expected to be quite relevant in the near term.

All innovation takes place with strong support from public policies. It is difficult to have a more favourable overall environment for fast-paced, adaptive and improvement innovations. Chinese firms already lead global innovation in internet distribution and sales, electronic games, and mobile internet. It is anticipated that they will do extremely well in online financial services. As most often is the case in Western companies, such momentum is not based on rocket science innovations, but in the Chinese environment, it is commercially effective.

This obsession with the market, aimed at providing sensible new offerings to discriminating consumers, is now moving on. This will not be a wholesale evolution. Chinese firms will progressively make innovation more effective (i.e. they will learn to deepen their level of innovations in design, services and technology creation to make their offerings more differentiated). They will do this while keeping their very sharp market-orientation.

Areas of high performance and excellence already exist, or will soon emerge and develop. They will spread progressively, but not evenly. After all, neither Europe nor the US have uniform regions or centres when it comes to entrepreneurship and wealth-creation.

This means that the regions that are currently the most vibrant, i.e. the Pearl River Delta (with Guangzhou, Shenzhen and Hong Kong) the BoHai Rim (Beijing-Tianjin-Hebei) and the Yangtze River Delta (Shanghai-Nanjing-Zhejiang) are likely going to lose their pre-eminence to the benefit of more inland cities and regions.

China as a global innovator

Barring major mishaps, China is poised to become a major source of innovation, not just for itself but also for the world. Keeping a strong controlling hand, the government will continue to gradually unleash and direct the energy of private firms.

Given the complexity of the processes at hand, but also of the speed at which China navigates change, following its own idiosyncrasies, the country is expected to become a global and more R&D-driven innovator across many sectors within *the next two decades*. At that point, its firms will likely be able to better leverage their engineering talent to impart new technical features into offerings and services, which will provide them with longer-lasting competitive advantages in the global markets.

It is very much hoped that, in the coming years, China will not hit too bumpy a road, as a result of a number of major challenges. One possibility is a major environmental crisis; others include: corruption, social unrest, an Asian-Pacific war caused for example by incidents over minuscule islands with Japan, one of China's historical enemies, and with whom no there has been no reconciliation. This is in sharp contrast with a wiser Europe, where Germany and France have reconciled. Another issue is the slowing down of the economy: hence again the need to innovate, which could be made worse by a financial crisis and/or the bursting of the real estate bubble.

Western firms must learn from China

The Western world has to become less ethnocentric and less self-centred. In this rambunctious, volatile environment, Western multinational companies have come to understand how to compete in China, particularly when it comes to technology-intensive innovation and adaptation of products, as well as the production of novel products for export outside China. A large number of them already have R&D presence in China, or are in the process of expanding it, making China a world centre for specific segments of the developments of new offerings.

Non-Chinese firms that want to learn from China hope to develop the more market-orientated, agile and rapid qualities of the Chinese environment. For close to two-thirds of them, foreign firms already judge that their Chinese competitors are either equally or more innovative than themselves.

Fear is not a good counsellor

When Europe regained its role as a major source of innovation and economic activity, after the Second World War, eventually to become what it is today (the largest and richest region in the world), everybody seemed to be quite content with the situation. In the 1950s, numerous US multinationals moved to Western Europe, to tap into both its talent and its markets. As a new world actor in innovation-intensive wealth creation, Chinese firms are about to play a similar role in Europe as US firms did several decades ago.

China will acquire a growing burden to act responsibly, both as an emerging source of innovation and as a global competitor. At the same time, China will provide a healthy stimulus to the non-Chinese world.

Rather than being seized by fear, the West must take up this stimulating challenge; it has many assets for doing so effectively. In fact, one of them is to learn the many lessons offered by the patterns of innovation *created in China*. For this, the Western world has to become less ethnocentric and less self-centred. China is on its way to becoming a global innovator. This is good for the country itself, and it is good for the world.

[1] www.tap4fun.com

[2] See Prof. Howard Yu, IMD 'Case Study on Xiaomi, No. 3-2410' (September 2013).

[3] See www.Haxlr8r.com.

[4] See www.mindray.com.

[5] During the final editing of the book, the Shanghai Stock Exchange collapsed by more than 30%, constituting this next crisis.

Bibliography

Beraha, Frédéric. *Apprendre de la Chine*. Paris. L'Harmattan, 2012.

Boutellier, Roman, Oliver Gassmann, and Max von Zedtwitz. *Managing Global Innovation*. Springer 3rd ed., 2008.

Breznitz, Dan and Michael Muphree. *Run of the Red Queen: government, innovation and globalisation*. Yale University Press, 2011.

Cannady, Cynthia. *Technology Licensing and Development Agreements*. Oxford University Press, 2013.

Chang, Gordon. *The coming collapse of China*. Random House, 2001.

Cheng, François. *L'éternité n'est pas de trop*. Albin Michel, 2002.

Chieng, André. *La pratique de la Chine*. Grasset, 2006.

Dodson, Bill. *China fast forward*. Wiley, 2012.

Dolla, Varaprasad. *Science and Technology in Contemporary China: Interrogating Policies and Progress*. Cambridge University Press, 2015.

Domenach, Jean-Luc. *Comprendre la Chine d'aujourd'hui*. Perrin, 2007.

Dyer, Geoff. *The contest of the century: the new era of competition with China*. Allen Lane, 2014.

Emmott, William. *Rivals*. Allen Lane, 2008.

Fannin, Rebecca. *Silicon Dragon: How China Is Winning the Tech Race*. McGraw-Hill, 2008.

Fernandes, Juan Antonio and Laurie Underwood. *China Entrepreneurs*. John Wiley Asia, 2009.

Florini, Ann M., Hairong Lai and Yeling Tan. *China Experiments: From Local Innovations to National Reform*. Brookings, 2012.

Fischer, William, Umberto Lago, and Fang Liu. *Reinventing Giants*. Jossey Bass, 2013.

Fu, Xiaolan. *China's path to innovation*. Cambridge University Press, 2015.

Grésillon, Gabriel. *Chine, le grand bond dans le brouillard*. Stock, 2015.

Haour, Georges. *Resolving the Innovation Paradox*. Palgrave MacMillan, 2004.

Haour, Georges and Laurent Miéville. *From Science to Business*. Palgrave MacMillan, 2011.

Hinyao, Wang and Suining Xu. *Chinese returnees: driving force of Chinese innovation*. People Publishing House, 2014.

Haihua, Zhang and Geoff Baker. *Think like Chinese*. The Federation Press, 2008.

Huff, Toby. *The Rise of Early Modern Science: Islam, China, and the West*. Cambridge University Press, 2003.

Jakobson, Linda. *Innovation with Chinese Characteristics: High-Tech Research in China*. Palgrave Macmillan, 2007.

Jolly, Dominique. *Stratégies d'entreprise en Chine*. Pearson, 2013.

Jolly, Dominique. *Le colosse aux pieds d'argile*. Maxima, 2014.

Keane, Michael. *Created in China: The great leap forward*. Routledge, 2007.

Le Belzic, Sébastien. *Quand la Chine vacillera, le monde tremblera: le cauchemar écologique.* Zinedi, 2003.

Lovell, Julia. *The Opium War.* Picador, 2011.

Lu, Giwen. *China's leap into the information age.* Oxford University Press, 2000.

Martin, Jacques. *When China rules the world: the end of the western world and the birth of a new global order.* Penguin, 2012.

Moyo, Dambisa. *Winners take all: China's race for resources and what it means for us.* Penguin, 2012.

Naisbitt, John and Doris Naisbitt. *Innovation in China: the Chengdu triangle.* Self-published, 2012.

Nea, Victor and Sonja Opper. *Capitalism from below.* Sage, 2012.

Needham, John. *Science and civilization in China.* Cambridge University Press, 1954.

Nie, Winter and William Dowel. *In the shadow of the dragon.* New York: Amacom, 2012.

OECD. *Reviews of national innovation policy China.* OECD, Paris, 2008.

Orcott, John. *Shaping China's Innovation Future.* Edward Elgar, 2010.

Petti, Claudio. *Technological Entrepreneurship in China.* Edward Elgar, 2012.

Phelps, Edmund. *Mass flourishing: how grassroots innovation created jobs, challenge and change.* Princeton, 2013.

Porter, Robin. *From Mao to market.* Columbia University Press, 2011.

Rein, Shaun. *The end of copycat China: the rise of creativity, innovation, and individualism in Asia.* Wiley & Sons, 2014.

Ricci, Matteo. *Journals on China in the XVIth Century.* Random House, New York, 1942.

Schramm, Ronald. *The Chinese Macroeconomy and Financial system: a US perspective.* Routledge, 2015.

Shambaugh, David. *China goes global.* Oxford University Press, 2013.

Sigurdson, Jon et al. *Technological superpower: China.* Edward Elgar, 2005.

Sieren, Frank. *Der China Schock: Wie Peking sich die Welt gefügig macht.* Ullstein, 2010.

Sun, Yifei, Max von Zedtwitz, and Denis Fred Simon. *Global R&D in China.* Routledge, 2008.

Tan, Yinglan. *Chinovation.* Wiley, 2011.

Temple, Robert. *The Genius of China: 3,000 Years of Science, Discovery, and Invention.* Inner Traditions, 2007.

Tse, Edward. *China's Disruptors.* Penguin, 2015.

Tzu, Sun. *The Art of War.* Nabla, 2010.

Vaitheeswaran, Vijay. *Need: Speed and Greed.* Harper, 2012

Van Someren, Taco and Shuhua van Someren-Wang. *Innovative China – Innovation race between East and West.* Berlin: Springer, 2013.

World Bank. *Promoting Enterprise-led innovation in China.* 2009.

Zeng, Ming and Peter Williamson. *Dragons at your door: how Chinese Cost Innovation is disrupting global competition.* Harvard Business School Press, 2007.

Zhang, Ying Ying and Zhou Yu. *The source of innovation in China.* Palgrave MacMillan, 2015.

Index